Leibniz and the Invention of Mathematical Transcendence

studia leibnitiana sonderhefte

Im Auftrage der Gottfried-Wilhelm-Leibniz-Gesellschaft e.V.
herausgegeben von Herbert Breger, Wenchao Li, Heinrich Schepers
und Wilhelm Totok †

In Verbindung mit Stefano di Bella, Francois Duchesneau, Michel Fichant,
Emily Grosholz, Nicholas Jolley, Klaus Erich Kaehler, Eberhard Knobloch,
Massimo Mugnai, Pauline Phemister, Hans Poser, Nicholas Rescher
und Catherine Wilson

Band 53

Michel Serfati

Leibniz and the Invention of
Mathematical Transcendence

 Franz Steiner Verlag

Bibliografische Information der Deutschen Nationalbibliothek:
Die Deutsche Nationalbibliothek verzeichnet diese Publikation in der Deutschen
Nationalbibliografie; detaillierte bibliografische Daten sind im Internet über
<http://dnb.d-nb.de> abrufbar.

© Franz Steiner Verlag, Stuttgart 2018
Satz: DTP + TEXT Eva Burri, Stuttgart
Druck: Offsetdruck Bokor, Bad Tölz
Gedruckt auf säurefreiem, alterungsbeständigem Papier.
Printed in Germany.
ISBN 978-3-515-12082-1 (Print)
ISBN 978-3-515-12083-8 (E-Book)

TABLE OF CONTENTS

LEIBNIZ AND THE INVENTION OF MATHEMATICAL TRANSCENDENCE. THE ADVENTURES OF AN IMPOSSIBLE INVENTORY

THE DISCOVERY OF THE TRANSCENDENCE

The invention of the mathematical transcendence in the seventeenth century is with good reason, linked to the name of Leibniz. He always claimed this creation – present in his work since 1673. Also one can only truly completely understand Leibniz as a mathematician, through the transcendence, and in relation to this, we can say that in some sense, he embodies it.

Descartes had created a completely new and wide symbolic frame (made of real polynomials with two variables) in which one considers plane curves. Far from reducing the field (as Hofmann wrongly claims it), this allowed the identification for the first time, of a certain concept of *all the curves* (i. e., all algebraic curves). Leibniz found and initially appreciated the Cartesian frame. But, as we shall see in this book, he was presented with, during his research, results and conceptualizations, gradually impossible to express in this context.

Initially, Leibniz qualified them with a vague word, the adjective '*transcendent*', most probably encountered in Nicholas of Cusa. Yet, he never associated this terminology to any philosophical – or theological – connotation; the term simply denoted what surpassed, 'exceeded' the Cartesian frame (it is the etymological meaning of 'transcend') without further constructive definition. Because, for Leibniz, all was first organized within the frame of the Cartesian mathematics, and in relation to it. It happened, however – it was in the nature of things – that what exceeded Descartes would later find some entirely new and diverse mathematical frames, both on a symbolical level (transcendent *expressions* and / or *functions*), a geometrical level (transcendent *curves*) and also a numerical level (transcendent *numbers*). There was only one term, but actually several concepts, which were not necessarily directly connected.

TRANSCENDENCE AND SYMBOLISM

On the symbolic level, the introduction by Newton of exponentials with fractional, then irrational exponents, had performed one of the first breaches in the Cartesian symbolism. In a symbolic approach that would 'transcend' both Descartes and Newton, Leibniz introduced then exponents with letters – unknown or indeterminate – of sign a^x, "exceeding any degree" (as he himself wrote). As we shall see, Leibniz granted too much hope and dedicated too much energy to them. It was what we call here the utopia of *letteralized exponentials* – coextensive with the hope of

Leibniz to have at his disposal an exhaustive inventory of this transcendent field that he brought to light.

A second breach in the Cartesian symbolism had been opened by the expressions of recently discovered functions, such as the logarithm or the exponential, then by obtaining, following Newton, quadratures *via* developments in power series (as $\sqrt{1+x}$). With his – remarkable – arithmetical quadrature of the circle, dated 1673, the young Leibniz, following Mercator, persevered victoriously in this direction by supplying a procedure – through the sum of an infinite series of rational numbers – in order to obtain what is today written as $\pi/4$ (which is a transcendent number).

TRANSCENDENCE AND GEOMETRY

On the geometrical level, the non-Cartesian curves caused a new problem of description for Leibniz; possibly even more complicated than that of symbolism: some well-known curves such as the 'logarithmic' could admittedly have an equation, but this one was not 'Cartesian' (*i. e.* it was not algebraic – in the modern sense). Others, such as the roulette or the spiral (both already rejected by Descartes) might have had a simple geometrical construction, but had no equation (in the Cartesian sense). Some others, such as the catenary or the trochoid, possessed a physical construction (mechanical, often using a thread), but without being provided however, with a Cartesian equation.

On the other hand, however, the introduction of the evolutes of plane curves, invented by Leibniz, following Huygens, allowed him – without significant effort – to mechanically construct transcendent curves from curves that were not transcendent. As it is detailed in this book, Leibniz was interested in examining closely all these eventualities. Leibniz was also the first to introduce the term of *algebraicity* – only with regards to curves – by forming a natural couple of opposites with *transcendence*.

LOOKING FOR AN INVENTORY

In this work, we deal at first with Leibniz's discovery, in the years 1673–1680, of different aspects of the transcendence that we have just mentioned. A decisive turning point took place around 1678. Indeed, in a first step, as we have said, Leibniz had used the term adjectively; he simply wanted to highlight – in diverse situations – what was not Cartesian. In a second step, however, he was not satisfied with the simple observation of various, scattered, established facts, but he adopted in his papers the word 'transcendence' in order to indicate, in its entirety, the territory as new as unknown, of the non-Cartesian entities. If the approach of denomination was positive, it referred to a negatively defined content.

Since the transcendence was the object of a denomination, it was natural that it arose for Leibniz (and incidentally, for John Bernoulli) the issue of its content, that is to say, the extension of the concept – in other words, of an *inventory of the transcendent*. One can understand to what extent this epistemological issue was nurtured by Leibniz's desire of a reasoned inventory of the non-Cartesian field. He wrote for example: "A method of invention would be perfect if we could predict, and even demonstrate, even before entering the subject, what are the ways by which we can reach its completion".

This approach was epistemologically natural, namely to inventory the non-Cartesian field in the same way as Descartes had inventoried the functions and the curves acceptable – according to his views. Retrospectively, it seems to us very naive today. This essay of inventory was indeed complicated by its various instances (symbolical, geometrical, numerical). However, above all, it was in fact made impossible by the completely negative character of the fundamental definition: is transcendent what is not Cartesian. Just like that of the irrational numbers (they are defined as non-rational) the absence of a positive definition of transcendent entities would, in reality, make impracticable any inventory. But, initially, Leibniz was not convinced of such impossibility. And we detail the – numerous and vain – attempts of inventory, in the later studies of Leibniz (1680–1690).

Leibniz's approach towards what he called definitively "the Calculus of Transcendent", however, became an essential part of the Leibnizian calculus. A very important article of Breger, dated 1986, 'Leibniz Einführung of Transzendenten'[1], will be a constant support in this study.

SYMBOLICAL INVENTORY?

In his attempts for a symbolical inventory, Leibniz first wanted to believe that the letteralized exponentials, of sign a^x, exhausted the subject. This is evidenced by numerous early texts. It was a very natural approach, directly following those of Descartes and Newton. Much later, his approach was entirely echoed by John Bernoulli, with his *percurrent calculus*. To a quite natural argument in favour of the exponential, rooted in an analogical symbolism (the 'primacy' of the exponential form), came another argument, significant in Leibniz's eyes: for him, these letteralized exponentials were *soluble* in the differential calculus – a point that we also analyse in detail.

Later, however, Leibniz realized that, so important may have been the creation of exponentials on the epistemological level, these did not fill all the universe of the non-Cartesian symbolism. The issue of knowing whether the percurrent calculus coincided or not with the transcendent calculus came to be therefore, in the forefront, between Leibniz and Bernoulli; in other words, whether or not, the a^x exhausted *all the content* of transcendent expressions. This caused a misunderstanding between both scholars that we will also later detail. More incidentally, Leibniz was also interested in 'intermediary' exponentials, of a particular type, which he called '*interscendent*', of sign $a^{\sqrt{2}}$ for example. These expressions were considered by their creator as 'less transcendent' than the previous ones, and this opened the

---

1 [Breger 1986].

door to a hierarchy in transcendence, which was pursued by following other approaches.

None of these considerations could however, be decisive: the new transcendent symbolism exceeded all these examples and all these instantiations.

Rather surprisingly, the approach of Leibniz found another type of symbolical completion within the calculus, in an article dated 1694, published in the Journal, des Savants, *Considérations sur la Différence qu'il y a entre l'Analyse Ordinaire et le Nouveau Calcul des Transcendantes* (*Considerations on the Difference Between the Ordinary Analysis and the New Calculus of the Transcendent*), a text of a considerable importance, which we also analyse. The quadratures were the geometrical support, and the sums, the symbolic support. Leibniz's discovery of a common structure to both foundational Triangles, the Arithmetical and the Harmonic, despite the fact that they were structurally opposed, and ultimately because of that very opposition, this led to the construction in the calculus of a harmony by reciprocity, that of the two assemblers of signs d and $\int$. For us, modern mathematicians, this did not (and could not) bring a definitive solution to the issue of the inventory, but, at least, and this time in a permanent way, symbolically constructed a new calculation scheme, based on the differential and integral calculus, and in which the harmony – so dear to Leibniz – was finally restored. In the new Leibnizian calculus, the summation of an arbitrary expression became from now on possible in all cases, the result being a transcendent expression.

GEOMETRICAL INVENTORY?

With regards to geometry, Leibniz began a precise analysis of the Cartesian conception of the 'geometry of the straight line – in order to criticize it. Remember that Descartes had decided to upset the Pappus' classification by completely reconstructing a typology governed by a new conceptualization, through the production of a pair of opposed concepts, specifically Cartesian: geometrical curves versus mechanical curves. In a natural way, Leibniz's criticism focused on the mode of production by Descartes of his geometrical curves. The latter indeed considered, *for every y fixed*, the solutions with respect to x of an equation of type:

$$a_0\,(y)\, x^n + a_1(y)\, x^{n-1} + \ldots + a_{n-1}(y)\, x + a_n\,(y) = 0 \;(= F\,(x,y))$$

where each a_j is a real polynomial (F is thus a real polynomial with two variables). This is a very particular conception, insufficiently analysed by commentators, in my opinion. However, as we explained it above, a large number of curves, commonly encountered at that time, escaped from the Cartesian analysis: first the curves that are now said parameterized, to which Leibniz delivered a short and brilliant study in the *Specimen Geometriæ Luciferæ* … But his two most original contributions to the Cartesian criticism were dedicated, on one hand to the evolutes, on the other hand to what he called the 'pointwise' constructions of the transcendent curves.

Regarding the first aspect, Leibniz noticed that the evolute of an algebraic curve is generally a transcendent curve. Therefore, he discovered a remarkable mathematical situation, because it is at the same time similar and reciprocal to that of quadratures: starting from algebraic curves, the new operation – the passage from one curve to its evolute – allows to construct, in a rather systematic way, transcendent curves, as does the squaring – the significant difference is that squaring lies in the field of sums, while the development falls within differentiation; that is to say, the two fundamental reciprocal operations invented by Leibniz.

The second aspect is perhaps less well known. In the early 1690s, Leibniz was interested in a category of transcendent curves that could be 'constructed', he said, by means of the ordinary geometry. The paradigm here was the catenary, namely a model of a transcendent curve to which we dedicate below a detailed study in the perspectives of Leibniz. Admittedly, the 'pointwise' construction of this one is made by means of the ordinary geometry, provided, however, one adds the additional required knowledge of a numerical quantity, namely the *transcendent number e*. Leibniz was quite aware of this. For him, knowing the value of *e,* was necessary and sufficient to construct the exponential curve, and thereby, in a second step, the catenary. It is thus this quantity, that Leibniz supposes immediately to be transcendent, which is the key according to him, and what he calls the 'construction of as many points as we want' of both curves by ordinary geometrical means.

On my part, I wanted to highlight the ontological difference, fundamental in my view, between these two statements: on one hand, 'points, as much as we want' and, on the other hand, 'an arbitrary point'. We already mentioned in this regard that Descartes rejected the quadratrix in the 'mechanical' curves, because it was not amenable to the construction of an arbitrary point of it.

I have therefore summarized the contributions of Leibniz in his search for symbolical inventories, as well as geometrical, and for *transcendent expressions,* as well as for *transcendent curves*. In this book, I also examine Leibniz's contributions to an adequate conceptualization of the *transcendent numbers*. However, one has to ascertain that they were hardly fruitful.

ON HIERARCHIES IN TRANSCENDENCE.
EXPLORATIONS BY REPRODUCTION

In the absence of an exhaustive inventory – bringing it to light appeared vainer day after day – geometers such as Leibniz and John Bernoulli dedicated themselves to a systematic and reasoned exploration, however this concerned only a part of the unknown corpus.

We have already noted above an instance concerning interscendent expressions. Another type of approach – more general and epistemologically significant – was to reproduce, by repeating it, a procedure to engender transcendent entities, which were considered more and more complex.

Trying to encompass the complexity of the unknown, through the organization of degrees in this complexity, is indeed a natural epistemological approach. If a procedure A is known to engender a transcendent $A(x)$ from an entity x (which is itself transcendent or algebraic), then, by applying A to $A(x)$, we produce a new transcendent $A(A(x))$ – it is a new entity that one will be tempted to consider as falling within a *second order of transcendence*; by organizing so an embryonic hierarchical scheme. One can observe the process on the geometrical level in Leibniz, with the successive evolutes: the evolute of an algebraic curve is (generally) transcendent; the evolute of the evolute is then a transcendent curve of second order, etc. We shall widely comment on these practices.

The approach was also dear to John Bernoulli, this time in the symbolic register. He implemented it in 1697, within the frame of his *degrees of percurrence*: since the production of the letteralized exponential a^x generates transcendent, then, for him, the repetition of the same procedure, which gives a^{x^y}, will spontaneously generate transcendent of order two, etc. Later, he attempts to resume the approach, still in the symbolic register; this time, he organizes *degrees of transcendence* through successive quadratures: the quadratrix $A(x)$ of an algebraic function x is (usually) a transcendent function. Under these conditions, the quadratrix $A(A(x))$ produces a transcendent of second order, etc.

RECEPTIONS OF THE TRANSCENDENCE

Introduced by Leibniz, the terminology of the transcendence was resumed by many of his correspondents, and by some influential mathematicians of the time, with the notable exception of Newton. The extreme diversity of protagonists' reactions however, reflects some embarrassment towards the strength and originality of Leibniz's developments; this perplexity was accompanying some hesitations by Leibniz himself regarding the exact, positive extension, of a concept negatively defined. In particular, we fully analyse an instructive controversy between Leibniz and Huygens – the latter even initially refused to accept the concept of transcendence.

Next, we continue the analysis of the reception of transcendence during the eighteenth and nineteenth centuries, both regarding the term and the concept, from Newton to Euler, Lambert, Liouville, and Hilbert. Note that Euler was the first who, in the eighteenth century, fully introduced the Leibnizian vocabulary of transcendence to the mathematical community, through his definitions for functions and curves; he pertinently analysed both aspects. On the other hand, his consideration of transcendent numbers was widely insufficient. Approximately at the same period, Lambert would give their first modern definition: an *algebraic* number is a root of an algebraic equation with integer coefficients; is *transcendent* a number which is not algebraic. We analyse in detail the reasons why Lambert's definition was *symbolically superior* to Euler's, and thus why it is the only one that has survived.

I also underlined that, if the modern concept did exist with Lambert – he was only interested in numbers – he did not always clearly use the term with this meaning. To refer to the modern concept he had uncovered, Lambert first used 'nombre au hasard' (number 'at random') and, concurrently, 'transcendent quantity' – but in fluctuating meanings.

As for numbers, the story of the crossed reception of the term and the concept of transcendence then presented very curious aspects. After Lambert and up till Hilbert, mathematicians would use, instead of 'transcendent number', the periphrasis 'number that is not reducible to algebraic irrationals'. It was only in 1900, in

Hilbert's works, that one could find the full coincidence of the term and the concept – for the numbers. Let us also note that, in the mid-nineteenth century (1844–1851), Liouville gave the first *demonstration* of the existence of transcendent numbers.

In the last part of the book, while studying diverse receptions of Leibniz's work, I dedicated a chapter to Auguste Comte, and to his philosophy of what he calls 'the transcendent analysis'. For Comte, there exists an objective scientific content, namely the transcendent analysis, which nevertheless received in history various successive subjective interpretations, mainly three in number, from Leibniz, Newton, and Lagrange. Next, Comte provides a reasoned history of 'the successive formation of the transcendent analysis', by examining the contributions of these three interpreters. As such, Leibniz is for him the genuine creator of the transcendence; but Comte expresses many reservations regarding the rigor of his process of creation. For him, Newton's conception was certainly more rigorous, however, not free from serious ontological weaknesses. As for Lagrange's approach, in which Comte sees 'the future' of the transcendence, he locates it in the necessary promotion of the 'abstraction' in mathematics – this point is essential for him. He rightly recognizes this aspect in Lagrange's conceptions. But if he insists on this point, he will nevertheless express other reservations on these – they are this time considered too abstract.

In a final chapter, I practiced a – very short – overview of the continuation of the story, namely the modern impacts of Leibniz's definitions of the couple of opposites: 'algebraic / transcendent'. At first, one must highlight this obvious fact: whether about numbers, functions or curves, the distinction is each time performed by characterizing *algebraic objects. Transcendent objects* are merely non-algebraic objects. Let me repeat: there is no positive property for characterizing the transcendent functions, namely no definition *for themselves*. Epistemologically speaking, the difficulty is thus obviously the same for demonstrating the irrationality or the transcendence; it is deep-rooted in the lack of definition of the object *per se*.

This rather exceptional situation in mathematics, has led, both for irrationality and transcendence, to use compelled proof methods, which are specific to these two domains, and to which the mathematician is absolutely forced.

FIRST PART
DISCOVERING TRANSCENDENCE

CHAPTER I
ON THE GERMINATIONS OF THE CONCEPT
OF 'TRANSCENDENCE'

The first occurrences, in Leibniz, of the term 'transcendent' are observed very early, during his stay in Paris. In Probst's works, "transcendent Curves" [Probst 2008a] (p. XXIX), and "algebraic and transcendent Curves" [Probst 2012] (p. XXXI), one finds numerous examples of early occurrences of this term. I examine here three of them: the first two, dated autumn 1673 and December 1674, the third one inside the first draft (they were two) of a letter to Oldenburg, sent on March 30th, 1675. However, the term does not appear any longer in the letter itself.

1673: "A TRANSCENDENT CURVE SQUARING THE CIRCLE …"

In December 1673, in *Progressio figurae segmentorum circuli aut ei sygnotae* (AVII, 3 N. 23, p. 266, l. 22), one finds probably one of the very first occurrences of the transcendent terminology. Leibniz dedicates himself first, (p. 266), to calculate $f(y) - f(y+1)$ for various values of f (the method is grounded on "the sum of all differences"; see below, Chap. IX). Finally, he chooses $f(y) = a^3/(a^2+y^2)$, and aligns the results. He then explains that these results serve to solve the quadrature of the circle, and to various general quadratures.

The text then displays two (mutually unconnected) occurrences of the word 'transcendent'. First, there is an obscure reference to sequences of integers which are not geometrical, that is to say, the ratio of two of their consecutive terms is not constant (as it is in 1, 2, 4, 8, 16, 32, etc.) but (indefinitely) increasing, as in 1, 2, 6, 24, 120, etc. Regarding this sequence, Leibniz says, "its dimension is transcendent":

> Inquirendum aliquando in naturas figurarum continue variantium, quae scilicet non sunt certae dimensionis sed transcendentes, ut: 1: 2: 6: 24: 120: huiusmodi figurae."

We find in this extract the first occurrence (without future occurrences) of a term deprived of any precise definition, where the word 'transcendent' is simply equivalent to 'exceeding'.

The term appears later in the text, in one of the very first of Leibniz's attempts to examine different methods of quadratures. For the young Leibniz, any quadrature was obtained *via* a quadratrix curve. This is indeed the pattern he established for squaring the circle (see below, Chap. II). As Leibniz writes (p. 267):

> "I see that this is similar, or is of the same nature as a transcendent curve squaring the circle and the hyperbola; and nobody has considered such curves in geometry, which it is necessary, if they are quadratrix, to describe by the continuous movement of some flexible cables. And they even can be said geometrical, although they are not liable to equations."

Let us note the emphasis on physical curves, which will have various extensions in Leibniz. This was the very first germ of the concept.

1674: ABOUT 'SECRET' GEOMETRY

Our second early text, dated December 1674, is devoted to the "secret" (*arcana*) geometry: *De progressionibus et geometria arcana et methodo tangentium inversa* (AVII, 3. N 39, p. 571, l. 6). The context is the research of curves given by a condition on their tangents (in other words, the inverse method of tangents), which is known to lead to a differential equation. The latter thus requires, *in fine*, a calculation of primitive curves of the equation, for which Leibniz first proposes a method where the quadrature leads to an algebraic result, and then judiciously expresses a reservation:

> If no analytical curve happens to succeed, the question would be to look if it is not possible to find a transcendent geometrical curve, as one of the kind of trochoids, evolutes, or helices.

This time, it is thus a question of a "transcendent curve", although it remains in a vague sense! Certainly, these curves are "mechanical" in Descartes' sense, but Leibniz does not even remark upon this. We then arrive to a text from January 1675, the importance of which is rightly underlined by Breger: Leibniz indeed begins to see the possibility of a more general concept of transcendence, but the word does not appear there. Breger writes[1]:

> One year later, Leibniz conceives an extension of the class of the curves, which are authorized in geometry. In an essay on quadrable curves, from January 1675[2], Leibniz defines such curves as analytical ones, for which the relation between ordinate and abscissa can be expressed by an equation. As what he calls "equation" is naturally an "algebraic equation", the analytical curves are thus algebraic curves (...); this essay, dated January 1675, is interesting because it deals with the question after the integration of the transcendent in geometry, while the word 'transcendent' itself never appears.

1675: THE LETTER TO OLDENBURG

The next main step takes place just three months after the previous essay, with the draft of the above letter to Oldenburg of March 30th, 1675 (A III, 1 N. 46, p. 201). This text is important, even if – let us repeat – it was not sent to the receiver. Leibniz – dealing with transcendent quantities – writes (p. 203, l.16) with the above-mentioned aim to establish an exhaustive classification of all kinds of quadratures:

> Analytical quadratures contain arithmetical quadratures, namely those by which the values of given magnitudes can be shown by means of either a certain number, rational or irrational, or even as a root of a certain equation, or even, of course, *via* some expression of the kind I used to call transcendent, when one demonstrates that it is impossible to find simpler ones.

---

1   [Breger 1986], 122.
2   'De Figuris Analyticis Figurae Analyticae Quadratricis Capacibus' = 'On Analytical Figures apt to be Quadratrices of Analytical Figures', AVII, 5, N.13, p. 99.

Looking more closely, the distinction in Leibniz between quadratures appears, in fact, somewhat unfeasible! Moreover, let us note that the term 'transcendent' is still used, here, in a vague manner ("some expression"). I do not pursue in detail here the contents of this draft, the study of which will be resumed below[3]. One then notes the significant occurrence of the term in a document dated April/June 1676. This is a Scholium[4] to *Quadraturae Circuli Arithmeticae Pars Prima*, itself a preparatory document to the project "Treatise on the Arithmetical Quadrature of the Circle". Leibniz there writes:

> As to the division in a given proportion of the angle, or of the ratio (…), this is definitely a transcendent problem, because it cannot be in fact reduced to any equation such as Vieta taught us to prepare, nor be built by means of curves such as those introduced by Descartes in geometry.

1678: THE "PERFECTION OF TRANSCENDENT CALCULUS"

We conclude this chapter with a last, important text. It will be our first comment about the letter from Leibniz to Tschirnhaus, sent in May 1678, from Hanover (GBM, 372–382). At that time, Leibniz's thought undoubtedly becomes more acute in relation to the concept of transcendence (the term appears frequently) and the typology of quadratures. First of all, this letter suggests interesting aspects of the philosophy of symbolism in Leibniz, which I will not discuss here[5]. Further on (*idem*, 376), Leibniz once more focuses on quadratures and on their typology; in the organization of the subject such as he conceives it, the quadratures of the circle and of the hyperbola are omnipresent as models (see Chap. II-A, below). In this text appear both the term 'transcendent' and also *letteralized*[6] exponents as a model of transcendence, actually one of their first occurrences in Leibniz. Indeed, he there deals – without any further restriction – with "transcendent quadratrices" and (elsewhere) with "transcendent quantities" about which, he writes, "the unknown enters the exponent" (*ibid.*, 376). Later, he mentions two (epistemologically central) hypotheses of choice as to exponential equations, and gives an early example of the *letteralization* of the exponent, with the equation:

$$x^y + \overline{ya^x} + etc = 0$$

Leibniz states:
The one who will be able to solve such (exponential) equations, either by transforming them into ordinary equations, or by demonstrating the impossibility of such a transformation, will be the one who will perfectly be able, either to find all quadratures, or demonstrate their impossibility, because all quadratures can be expressed by such transcendent equations. That's why we

3   Also, see an analysis of this draft in [Hofmann 1974], 127–128.
4   AVII, 6. N. 20. Here page 219.
5   See [Serfati 2005], 269–319, and [Serfati 2010b].
6   This term is my own neologism. It characterizes the systematic substitution of a letter instead of a digit.

must improve this kind of transcendent calculation, through which we shall absolutely have all we seek in this field (*ibid.,* 377).

Let us underline that if the above equation appears – even today – complex, as to its algebraic solution, it will be – on the other hand – for Leibniz, directly amenable to his differential calculus, (see Chap. IV. D5, below). After 1678, both the terms 'transcendent' and 'transcendence' become conventional in Leibniz's mathematical writings.

CHAPTER II
SQUARING THE CIRCLE

II-A THE ARITHMETICAL QUADRATURE OF THE CIRCLE (1673), OR THE VERY FIRST MATHEMATICAL GLORY OF LEIBNIZ

The quadrature of the hyperbola in Mercator

We have just seen the importance of quadratures for Leibniz during this initial phase. This question had been fundamental since antiquity, originating with the quadrature of the parabola (Archimedes). However, some new and important results appeared in the second half of the seventeenth century, initially with the quadrature of the hyperbola, carried out by Mercator[1] in his *Logarithmotechnia* (1668), which had impressed Leibniz[2].

We briefly summarize, in modern terms, Mercator's method. It consists of calculating $\int \frac{dx}{1+x}$ (i.e., the area under the curve of equation $y = \frac{1}{1+x}$) by expansion in Taylor series:

$$\frac{1}{1+x} = \sum_{n \geq 0} (-1)^n x^n \ (1)$$

He calls this formula, the 'long division'. Afterwards, he integrates, term-by-term (the quadratures of the various x^n, called *paraboloids*, were known at the time[3]). Mercator finally finds $\sum_{n \geq 0} (-1)^n \frac{x^{n+1}}{n+1}$. The link between this sum and ln $(1+x)$ had also been well known since Grégoire de Saint Vincent. For the sector of the hyperbola $(0 \leq x \leq 1)$, we thus obtain:

$$\sum_{n \geq 0} (-1)^n \frac{1}{(n+1)} = 1 - \frac{1}{2} + \frac{1}{3} - \frac{1}{4} + \text{etc.}$$

That is, in modern terms, we obtain the sum of the alternating harmonic series (recall that it is conditionally convergent)[4].

---

1   [Mercator 1668].
2   GM V, 383.
3   On the quadrature of paraboloids, see, for instance, [Parmentier 2004],102–103 (and Leibniz's comment). See also Leibniz's letter to Wolff, dated 1713 (GM V, 384).
4   Note that the term-by-term integration of the Taylor series (1) is still valid for x = 1, even if the series is there only conditionally convergent (this result is usually called 'Abel's second theorem').

Quadratrices and symbolic substitutions:
the mathematical ideas of Leibniz

Leibniz's method for squaring the circle was a remarkable adaptation of that of Mercator, *via* a geometric-differential procedure inspired by Pascal (the 'character-istic triangle'). Leibniz replaced, first of all, the area sought with that of some cir-cular sector (i. e., the area obtained by joining a fixed point O of the circle to two other points of the circle, one being fixed, the other variable). He then replaced the same area by that situated *under* the *irrational* algebraic curve of equation $x = \sqrt{\dfrac{1-y}{y}}$. Leibniz immediately transforms the latter curve – which he calls the 'quadratrix of the circle'– by reciprocity into the rational curve of equation $y = \dfrac{1}{1+x^2}$; the success of this method brought about his long-lasting interest in curves as quadratrices[5]. One can show that this calculation of the area of a circular sector covers exactly, in modern terms, the formula for the line integral, as is now well known as:

$$\frac{1}{2}\int (x\,dy - y\,dx)$$

In a second step, the method of Leibniz consisted (basically) of replacing x by x^2 in relation (1) of Mercator, and then in integrating, term-by-term, the result between 0 and 1, to thus find:

$$\sum_{n\geq 0}(-1)^n \frac{1}{2n+1} = 1 - \frac{1}{3} + \frac{1}{5} - \frac{1}{7} + etc.$$

This gives the value of the ratio of the area to the square of the diameter (in modern terms $\frac{\pi}{4}$). For us, the substitution ($x \to x^2$) goes without saying (!); however, it was, at that time, profoundly new. One can hardly imagine today the sum of the conceptual difficulties met by the scientific minds of the time, still impregnated with 'concrete' geometrical truths, to conceive such a substitution involving only symbolic material rather than the geometrical substances of the curves concerned – namely, hyperbola and circle – which had been well known for a long time. The substitution made by Leibniz was, indeed, remarkable. It must clearly be credited to this author as one of his claims to glory – it was, to my knowledge, the first histori-cal example of effective operational substitution in the calculus. Moreover, it is

5   We give here only the scheme of the method. In modern terms, Leibniz's calculation actually leads to: $x = \sqrt{\dfrac{y}{2-y}}$, or else $y = \dfrac{2x^2}{1+x^2}$. But $\dfrac{2x^2}{1+x^2} = 2\left(1 - \dfrac{1}{1+x^2}\right)$. Thus, we are returned to the Taylor series expansion of $\dfrac{1}{1+x^2}$ On the detail of the proof, see, for example, [Hofmann 1974], 59–60.

hardly necessary to emphasize how much this type of procedure requires an appeal to symbolic writing.

Throughout his life, Leibniz was rightly proud with this result, which marked his authentic entrance into a mathematical career, and on which Huygens, whose judgment mattered for him so much, immediately congratulated him greatly.

Before Mercator and Leibniz, the quadratures of the circle and the hyperbola had established insuperable obstacles. Nonetheless, it is possible that we measure wrongly, nowadays, the meta-theoretical importance of such a result as the Leibnizian quadrature. It says something precise on what was, at the time, a thousand-year-old mystery[6], even if this precision was not primitively designed as such by its author: this procedure would then have resulted in a number, which would definitely have given a value, but Leibniz, initially, refused this conception (see Chap. VII *infra*). Admittedly, Wallis and Brouncker had previously given a certain approach to the same mystery, in the form of an infinite product for the former (in modern terms $\dfrac{\pi}{2} = \prod_{n=1}^{\infty} \dfrac{(2n)(2n)}{(2n-1)(2n+1)}$)[7], or of a continued fraction expansion[8] for the latter. However, Leibniz's result imposed on everyone, and first in the eyes of his author himself through its arithmetical implicity and harmony[9]. This issue is well known.

Leibniz on "the exact proportion …"

I give, here, some brief historical supplements. Leibniz had established his result in 1673 by home-made geometric means. He obviously had intended to publish it, and he left, upon leaving Paris, a manuscript in this purpose. However, the representative of the publication did not exhibit much enthusiasm; even so, Leibniz, almost immediately afterwards, was to work out the general theory of his differential calculus, into which as an imperative he wanted to integrate – having redrafting it – his *Memoire* on the quadrature.

He thus refrained temporarily from publishing, wishing to have time for this reorganization; and it was only in 1682 on the occasion of the foundation in Leipzig of the scholarly journal *Acta Eruditorum*, that he decided to publish it in this jour-

---

6   On the detailed history of the research into the quadrature of the circle, see [Montucla 1754] and [Serfati 1992].

7   In *Arithmetica Infinitorum* (1655).

8   Namely $\dfrac{\pi}{4} = \cfrac{1}{1+\cfrac{1^2}{2+\cfrac{3^2}{2+\cfrac{5^2}{\text{etc.}}}}}$. This formula had been conjectured by Brouncker (ca. 1660).

9   Leibniz states (GM V, 120): "Taken in its totality, the series thus expresses the exact value. And although we cannot write the sum in a single number, and it continues to infinity, the mind can conceive it properly in its *entirety*, insofar as it is constituted by a single law of progression."

nal; this was the *De Vera Proportione* (…)[10]. This text contains a remarkable explanatory memorandum but strangely, no demonstration of the central result. However, Leibniz also drafted many sketches and documents of the proof (see GM V [11]).

The importance of his result on the arithmetical quadrature of the circle led Leibniz to such a line of questioning: admittedly, it provides a value, but is this value a number? What is the probative value of this type of quadrature? What are, given these conditions, the diverse possible types of quadratures? Are they all equally legitimate? These were questions that nobody, really, had raised before Leibniz, and which were directly inspired by the form of his result. All this concludes with him in a new and remarkable epistemological approach, an inventory and a typology of quadratures. They are found in many early texts – already mentioned in the previous chapter. It is thus within the context of an inventory of quadratures, whether analytical or geometric – or even mechanical – that he wanted to identify the place of his arithmetical quadrature and its transcendent expressions.

Some reflections on the integration of rational fractions

We can, in modern terms, summarize some of our conclusions by underlining that the issue of the quadrature of rational fractions was almost certainly instructive for Leibniz. Rational functions are very common and very simple, by a degree only more complex than polynomials. However, the sum (= quadrature) of a rational fraction is *not* a rational fraction, except in a small number of exceptional cases. The question is more widely developed *infra* in (IX-D). In other words, the summation of an expression of this type (written in the usual symbolic system) cannot generally be made in the same system. What about, for example, these three summations?

$$\int \frac{1}{1+x} \text{ or } \int \frac{1}{1+x^2} \text{ or } \int \frac{1-x}{x^3+x-1}?$$

Yet, as we have just seen, Leibniz had every reason for knowing that the first two examples correspond in geometry to two crucial problems of quadrature of the hyperbola and the circle, respectively. Furthermore, as we have also seen, those values are given *via* the sums of infinite numerical series obtained by the term-by-term integration of some power series. Consequently, the quadrature methods, and thus those of summations, fall under an inescapable infinitary frame – that of a new Analysis. Leibniz then took advantage of this situation to set up a new conceptualization, that of a *transcendent expression*. Accordingly, we understand the impor-

---

10 *De Vera Proportione Circuli Ad Quadratum Circumscriptum in Numeris Rationalibus Expressa* (On the exact proportion between a circle and its circumscribed square, expressed in rational numbers).

11 See, for instance, *Praefatio Opusculi de Quadratura Circuli Arithmetica* (GM V, 93) and *Compendium Quadraturae Arithmeticae* (GM V, 99).

tance of quadratures in the genesis of the concept of transcendence. These aspects are more extensively developed *infra* Chap. IX.

II-B LEIBNIZ AND THE IMPOSSIBILITY
OF THE ANALYTICAL QUADRATURE OF THE CIRCLE OF GREGORY

The true quadrature of the circle and hyperbola,
by James Gregory

Another major source of reflection on transcendence among Leibniz's contemporaries was certainly the famous and violent controversy between James Gregory and Huygens on the impossibility of the analytical quadrature of the circle, following the publication in 1667 of Gregory's *Vera Circuli et Hyperbolae Quadratura*. Leibniz had knowledge of it in late December 1673, *via* Huygens who had lent him the book, and on which he took notes (this point is noted by Breger[12]). Doubtless Huygens granted this loan to Leibniz so that the latter, unable to rule on the (painful) controversy, could come provide grist to his mill – that is to say, to supply him with some supplementary arguments in order to refute Gregory. Thus, in 1673 (a period contemporaneous to his discovery of the *arithmetical* quadrature of the circle), Leibniz was confronted with the hypothesis of Gregory concerning the impossibility of an *algebraic* quadrature of the circle [13].

The quarrel was met with a widespread echoing among his contemporaries. A century later, in his book *Histoire des recherches sur la Quadrature du Cercle*, Montucla [14] explains at length and with great care the demonstration of Gregory. It remains today the subject of abundant commentary (examples include Whiteside[15], Dijksterhuis[16] and Hofmann[17]).

In the *Vera Circuli*, Gregory intended to rigorously demonstrate that the analytical quadrature of the circle is impossible. The aim was certainly laudable (because the result is exact ...) but the demonstration was insufficient and – ultimately – wrong. We understand today, in retrospect, that it needed the mathematical tools and more elaborate concepts of numerical transcendence, which Gregory could hardly have conceived of at that time[18]. Huygens and Leibniz were both cer-

---

12　See [Breger 1986], 15.
13　According to [Hofmann 1976], Leibniz had visited Huygens the day before New Year's Eve in 1673, apparently to wish him a happy New Year. He probably told him about his discoveries on the arithmetic quadrature of the circle. This had prompted Huygens to lend him various works by Gregory, including the *Vera circuli*.
14　[Montucla 1754], especially 90–91.
15　[Whiteside 1961], 226–227 & 252–253 & 270.
16　[Dijksterhuis 1939], especially 481.
17　See [Hofmann 1974], pages 63-64, which contains a good analysis of Gregory's method. See also [Hofmann 1976], p. LIV–LVI.
18　The impossibility of the analytical quadrature of the circle was established in 1882 in an article by Lindemann on the transcendence of π ([Lindemann1882]), following the landmark work of

tainly persuaded of the insufficiency of the proof, but this insufficiency was hardly easy to prove. Nonetheless, they tried (unsuccessfully) to convince Gregory of it.

Gregory's 'proof' nevertheless collects elements of mathematical analysis which remain today relevant and which were indisputably very interesting for the young Leibniz (cf. *infra*). First, one finds the idea of the construction by recurrence of two sequences (what Leibniz later called 'two *series replicatæ*')[19], an entirely new process at the time. In a second step, the convergence of these two *adjacent* sequences (in the modern sense) to a same limit, and therefore a statement (on the model of what later would be the 'Cauchy condition for convergence of sequences') about the existence of a limit without knowing its value in advance – for the first time in the history of mathematics. This was what will be called below *the pattern of geometric–harmonic mean* (G-H). In modern terms:

$$a_{n+1} = \sqrt{a_n b_n} \ (\underline{G}) \text{ and } \frac{2}{b_{n+1}} = \frac{1}{a_{n+1}} + \frac{1}{b_n} \ (\underline{H})$$

This first part appears as interesting and correct to the contemporary reader. We give later on in this chapter a brief proof in modern terms (*The conception of the geometric–harmonic mean in Gregory*). Leibniz, who well understood the idea, would develop it and would even succeed in generalizing it (see below Leibniz's *"means by composition"*).

The second phase of the *Vera Circuli* is highly problematic. Indeed, for a second time Gregory sought to demonstrate that the common limit of the two sequences thus obtained is not 'Cartesian' – that is to say, is not algebraic – in the sense that it cannot be written algebraically with respect to the initial data. As such, it is really the first attempt to prove that π is a transcendent number, in the modern sense. I would insist, even if his proof is not correct, that Gregory very closely approached the *modern issue* of the transcendence of numbers. It will, however, be necessary to wait for Lambert before we gain a truly appropriate definition of the transcendent numbers (see *infra* Chap. XIII-B).

Leibniz, for his part, did not doubt the veracity of the result, namely the impossibility of the analytical quadrature – but like Huygens[20], he was not convinced by the 'proof' offered by Gregory[21]. What is important for us here is that he did not

---

Hermite on the transcendence of *e* by [Hermite 1873a]. See below, Section VI-D, as well as our detailed study [Serfati 1992],117–162. See also [Waldschmidt 1983] and [Lebesgue 1932].

19   Cf. [Probst 2006]: "Instead of Gregory's double convergent sequences, Leibniz favors the research for convergent sequences which go by pairs (*series replicatae*) generally (A III, 1. N. 39, 64, 68), of which Gregory's sequences represent specific examples. As regards the concept of convergent sequences in Gregory, Leibniz criticized the nature of the double sequence, since usually (in the modern sense of convergent) a single sequence is already sufficient."

20   Dijksterhuis points out (p. 483): "Huygens denies the right to assume that the common limit t of the two series a_1, a_2,..., and b_1, b_2... if it exists, can always be expressed analytically in a_i and b_i. Gregory never succeeded in refuting this argument."

21   Breger notes: "Leibniz was persuaded that Gregory's proof was incomplete, but he was already convinced (according to Hofmann) that the algebraic quadrature of the circle was impossible", [Breger 1986], 121–122.

question the validity of the result – that is, and in other words, the non-algebraicity of π. At this time, however, Leibniz would not pursue Gregory's approach in seeking some proof of numerical non-algebraicity.

The geometric–harmonic mean of Gregory

In modern terms, one can present Gregory's approach as follows: given a sector built from the centre of a conic, Gregory considers a triangle inscribed in this sector and a quadrilateral circumscribed to it with areas a and b, respectively $(a<b)$[22]. The value of the required area of the sector lies between a and b. Gregory then replaces the couple (a, b) by the couple (a_1, b_1) constituted, on the one hand, by their geometric mean a_1:

$$a_1 = \sqrt{ab} \ (\underline{G}_1)$$

and, on the other hand, by the harmonic mean, say b_1, of b and a_1:

$$\frac{2}{b_1} = \frac{1}{a_1} + \frac{1}{b}, \ i.\ e.\ b_1 = \frac{2a_1 b}{a_1 + b} (\underline{H}_1).$$

He shows that, in the case of the circle or the ellipse, we have:

$$a < a_1 < b_1 < b$$

(in the case of the hyperbola, the sense of the inequalities is reversed). Afterwards, he resumes the same procedure with a_2 = Geom (a_1, b_1) and b_2 = Harmon (a_2, b_1), and so he builds by recurrence a double sequence (a_n, b_n), satisfying the conditions $(\underline{G})$ and $(\underline{H})$ above. Next, he shows (according to Whiteside) that they constitute what we call today 'two adjacent sequences', namely:

$$a_n \text{ increasing}, b_n \text{ decreasing, and } \lim (b_n - a_n) = 0$$

Gregory concludes that these are two *convergent* sequences (he introduced the term into mathematics[23]), and that they admit a common limit t, which is the value of the required area of the sector (I give below a modern direct proof of this result at the end of this chapter[24]).

What Huygens and Leibniz after him will refute is not the existence of such a geometric-harmonic mean, but the proof of the subsequent conclusion derived by Gregory, namely: t, considered as a function of *a* and *b*, cannot be algebraic, in the

---

22   See a geometrical descriptions in [Whiteside 1961], 226 and [Hofmann 1974], 65.
23   Thus, the term 'convergent' was initially introduced by Gregory *only to indicate a couple of sequences* (in the plural, thus) with the connotation of 'confluentes', or else 'which aim towards the same end' – that is to say, according to the meaning of the word in the common language, for example, "convergent reasons to decide on a choice." The modern terminology has fully concealed this conception, so as to build the concept of convergence of a (single) sequence.
24   It shortens the proof of [Whiteside 1961], 266.

following sense – there is no function F, rational or radical irrational, such that F (t) = 0. Such was the object of a fierce controversy.

Leibniz and the "convergence" of adjacent sequences (1676)

Leibniz was very interested in the construction of Gregory's (G-H) mean. Three years after the previous episode, he resumed the subject with a remarkable attempt of extension, in a manuscript dated June 1676: *Series convergentes duae*[25]. He there dealt, in some detail, with the construction of couples of sequences defined by recurrence. If the process is now well known and regularly taught, it was surprisingly new at the time. In this manuscript, Leibniz seeks especially to capture the essence of Gregory's method in order to generalize it. We propose below a brief analysis[26].

Leibniz first takes, for example, the arithmetic-harmonic mean (hereafter (A-H)) rather than Gregory's (G-H). He begins by reconstructing repeated sequences in the style of Gregory:

a and b are given numbers. Let A = Arith (a, b) and B = Harmon (a, b).

He then defines (A) = Arith (A, B) and (B) = Harmon (A, B), etc.
In modern terms:[27]

$$a_{n+1} = \frac{a_n + b_n}{2} \text{ and } b_{n+1} = \frac{2a_n b_n}{a_n + b_n} \text{ (relations (A-H))}$$

If the difference between the two terms a_n and b_n becomes smaller than any given quantity, then, Leibniz writes, there is an ultimate term of the convergence, which we denote by t.

Moreover, in the case (A-H) envisaged by Leibniz, we can, as he himself points out (p. 799, l. 17), calculate the value of the limit. There is indeed an *invariant for the couple of transformations* (A) and (H), namely an (algebraic) continuous function ϕ, such that:

$$\phi(a_n, b_n) = \phi(a_{n-1}, b_{n-1}) \text{ for every n}$$

If such an invariant exists, we have, *in fine*, $\phi(a_n, b_n) = \phi(a, b)$, and, passing to the limit (it exists, and ϕ is continuous), then $\phi(t, t) = \phi(a, b)$, which provides an algebraic equation to calculate t.

---

25   AVII, 3. N. 64, p. 800, l. 12.
26   The term 'transcendent' appears, however, in this text, only in an anecdotal way, via some hypothetical letteralized exponential t^w (cf. *infra*).
27   As compared to Gregory's relations (G) and (H) above, one notes, however, some difference. In Leibniz, the value of the couple $a_{n+1} b_{n+1}$ at rank n+1 is given according to the value of the same couple at rank n:

$$a_{n+1} = f(a_n, b_n) \text{ and } b_{n+1} = g(a_n, b_n)$$

Gregory gives the value of the same couple in mixed way, according to its values at both ranks n and n+1:

$$a_{n+1} = f(a_n, b_n) \text{ and } b_{n+1} = g(a_{n+1}, b_n)$$

The conception offered by Leibniz is the more popular today.

In the case of (<u>A-H</u>), the function ϕ defined by $\phi(x,y)=\sqrt{xy}$ is suitable as an invariant, since we have:

$$\sqrt{a_{n+1}b_{n+1}}=\sqrt{a_n b_n}=\ldots=\sqrt{ab}.$$

Therefore, at the limit, $\sqrt{t \cdot t}=\sqrt{a \cdot b}$ and $t=\sqrt{ab}$. This is the arithmetic-harmonic mean – it coincides with the geometric mean – and it is thus algebraic irrational with respect to a and b.

Leibniz's "means by composition"

Leibniz thus says that, instead of A = Arithm and B = Harmon, he may as well choose two other functions A and B as sources of iteration. Under the condition that the obtained sequences are adjacent, he postulates that we always have, at the limit, a way to calculate the value of t. He chooses as an example:

$$A = \text{Geom (a, b)} \quad \text{and} \quad B = \frac{2ab}{a+\sqrt{ab}}$$

As we have seen, Leibniz always wants to have relations such as $a_{n+1}=f(a_n,b_n)$ and $b_{n+1}=g(a_n,b_n)$. The functions f and g are, however, restricted to be continuous and to satisfy $f(a,a)=a$ and $g(a,a)=a$, for all a^{28}: we will say in the sequel that f and g are ***mean-functions*** (m-functions, for short).

Note that the set of all m-functions forms a real algebra of functions M^{29} which is, furthermore, *stable by composition* in the following sense: if f, g and h are any three elements of M, then the function K of two real variables defined by K = h (f, g), that is:

$$K(x,y)= h(f(x,y), g(x,y))$$

also belongs to M^{30}. On the other hand, if a_n and b_n are adjacent sequences, they both converge to the same limit x. By continuity, f and g admit x for a fixed point; that is to say, they must verify:

$$f(x,x) = x \quad \text{and} \quad g(x,x) = x.$$

Yet this is naturally true, *by structure*, if f and g are *mean-functions*, whether arithmetic, geometric or harmonic[31]. One thus obtains three types of means by composition, (<u>A-H</u>), (<u>G-H</u>) and (<u>A-G</u>)[32]. But it is worth noting that this property is still true with the new Leibnizian example B, above. Indeed, we have:

---

28   This is due to the geometrical origin of the problem concerned.
29   Let us note $E = IR^{(IR^2)}$, the algebra of all the real functions of two real variables, and M = {f ∈ E / f(x,x)=x}. Any element of M is called a ***mean-function***. M is clearly a sub-algebra of E.
30   One has $K(a,a)= h(f(a,a), g(a,a)) = h(a,a) = a$.
31   Or any other m-function. For instance, any mean of order p.
32   The study of the arithmetic-geometric mean (<u>A–G</u>) is a classic today.

$$B(a,a) = \frac{2a^2}{a + \sqrt{a^2}} = \frac{2a^2}{2a} = a$$

This example is a direct application of our conclusion *supra* on the stability by composition[33] of the set of all the mean-functions. This design was, therefore, due to Leibniz[34], who thus showed the extent to which he had analysed Gregory's mechanism. I called the functions resulting from this process the *means by composition of Leibniz*.

On the other hand, in the arithmetic-harmonic case above, (A-H), there is an invariant ϕ which allows the calculation of the limit. Moreover, Leibniz believed – in my opinion – that this property was valid in all other cases of means by composition that he considered. In general, he thought that, by choosing repeated sequences, one can always get an invariance φ as a function, but the latter may optionally be transcendent (not necessarily algebraic). Thus, he says (p. 801), if we have ϕ (a,b) = $a^\omega + b$, then ultimately, at the limit, $t^\omega + t = a^\omega + b$; this is a standard transcendent equation with regard to the unknown t, whose second term is known, if we know omega[35]. Thus, the limit t does exist, but it is not algebraic.

In this passage of *Series convergentes duae*, Leibniz indeed returns to criticize Gregory's *final* argument in the *Vera Circuli*, the one that aimed to demonstrate the impossibility of the analytical quadrature. For Gregory, the impossibility of any algebraic expression for the area of the sector leans explicitly on the consideration of the non-existence of such algebraic invariants.

I provide a sketch below of a brief commentary on a complex process which the reader will find developed in more detail in the literature[36]. First, if there is such an algebraic invariant ϕ, then, as we have just seen in Leibniz, the relation $\phi(t,t) = \phi(a,b)$ provides the area t of the sector by an algebraic equation. The area is, therefore, algebraic, in the modern sense. Hence, it is the *impossibility* of any such relation that Gregory tried, it seems, to demonstrate. It is upon this point that his argument is still largely insufficient, even if his approach is interesting[37]. We therefore emphasize, once again, that Leibniz had fully understood Gregory's complex process[38]. We now give a modern proof of Gregory's result.

---

33 One indeed has the relation of composition B(a,b) = h(f,g), where f, g and h are three m-functions, namely: h(a,b) = Harmon (a,b), f(a,b) = a (f is a m-function) and g(a,b) = Geom (a,b).

34 To my knowledge, we find this conception nowhere other than in Leibniz.

35 "Exhibente terminationem, videtur semper æquatio transcendens inveniri posse (...)" (p. 800).

36 See [Whiteside 1961] (p. 270), [Hofmann 1974] (p. 65–6), [Dijksterhuis 1939] (p. 482–483).

37 See [Hofmann 1974] (p. 66): "The idea is truly ingenious, especially when one remembers the primitive tools at Gregory's disposal."

38 In the aforesaid draft of his letter to Oldenbourg (dated March 30th, 1675), Leibniz pays a glowing tribute to Gregory's mathematical ideas (A III, 1. N. 46, p. 204).

A mathematical appendix:
the geometric-harmonic mean (G-H)

Theorem (Gregory): Let a and b be two real numbers such that $0 < b < a$. One defines $a_0 = a$ and $b_0 = b$. And, for every $n \geq 0$:

$$b_{n+1} = \sqrt{a_n b_n} \ (\text{G})$$

$$\frac{2}{a_{n+1}} = \frac{1}{b_{n+1}} + \frac{1}{a_n} \ (\text{H})$$

Both sequences a_n and b_n converge to the same real number L, called the *geometric-harmonic mean* of a and b.

Proof. We first show by an immediate induction that a_n and b_n, defined above, are positive real numbers. The two sequences are therefore well defined.

Lemma 1. One has, for every $n \geq 0$: $0 < b_n < b_{n+1} < a_{n+1} < a_n$ (R_n).

Proof. One has $(b_1)^2 = a.b$ (by (G)). But $b^2 < ab < a^2$. Hence, $b < b_1 < a$.

On the other hand, $a_1 = \dfrac{2b_1 a}{b_1 + a}$, by (H) and $a_1 - b_1 = \dfrac{2b_1 a}{b_1 + a} - b_1 = \dfrac{b_1 a - b_1^2}{b_1 + a} = b_1 \dfrac{a - b_1}{b_1 + a} > 0$.

Hence, $0 < b < b_1 < a_1 < a$.

Take for the induction hypothesis:

$$0 < b_n < b_{n+1} < a_{n+1} < a_n \ (R_n)$$

One has:

$$(b_{n+1})^2 < a_{n+1}.b_{n+1} < (a_{n+1})^2.$$

Since, by (G), one has: $(b_{n+2})^2 = a_{n+1} b_{n+1}$ then $b_{n+1} < b_{n+2} < a_{n+1}$.

Moreover, by (H):

$$a_{n+2} = \frac{2b_{n+2} a_{n+1}}{b_{n+2} + a_{n+1}}.$$

Hence, $a_{n+2} - b_{n+2} = \dfrac{2b_{n+2} a_{n+1}}{b_{n+2} + a_{n+1}} - b_{n+2} = \dfrac{b_{n+2}(a_{n+1} - b_{n+2})}{b_{n+2} + a_{n+1}} > 0$. It results that $a_{n+2} > b_{n+2}$. So, (R_{n+1}) is valid. Hence, (R_n) is valid for every n. The sequence b_n is therefore increasing and bounded above; in the same way, a_n is decreasing and bounded below. Both sequences thus converge to L and L', respectively, with $0 < L \leq L'$. We then prove that $L = L'$.

Lemma 2. One has the relation:

$$a_{n+1} = \frac{2b_{n+1}^2}{b_n + b_{n+1}} \ (1)$$

Proof. By (<u>H</u>), one has $a_{n+1} = \dfrac{2b_{n+1}a_n}{b_{n+1}+a_n}$. By multiplying the numerator and the de-

nominator by b_n, one gets $a_{n+1} = \dfrac{b_n(2b_{n+1}a_n)}{b_n(b_{n+1}+a_n)} = \dfrac{2b_{n+1}^3}{b_n b_{n+1}+b_{n+1}^2} = \dfrac{2b_{n+1}^2}{b_n+b_{n+1}}$, from which

relation (1) follows.

The two sequences a_n and b_n converge to L and L' respectively. Taking the lim-

its in (1), one then gets $L' = \dfrac{2L^2}{2L} = L > 0$ [39].

---

39   [Whiteside 1961], p. 266, gives for this theorem an interesting, but much longer and more
     complicated, demonstration. Nonetheless, his proof follows, he says, that of Gregory, by dem-
     onstrating as to b_n (for example) the following geometric inequality (established, he says, by
     Gregory):
$$(b_{n+2} - b_{n+1}) < \frac{1}{4}(b_{n+1} - b_n)$$

CHAPTER III
THE POWER OF SYMBOLISM:
EXPONENTIALS WITH LETTERS

So far, we have recognized in Leibniz a main source of transcendence in quadratures, with its origin in his Arithmetic Quadrature of the Circle. It essentially consisted of *transcendent curves*. In this chapter, we examine the origin of transcendence of another nature, through *exponentials with letters*. This time, the constitution is symbolic. It consists of transcendent expressions (*i. e.* for us, functions). We have already noted such an instance, in the letter to Tschirnhaus of May 1678. I shall describe their genesis, starting from two letters, sent two years earlier (1676) from Newton to Leibniz, both *via* Oldenburg[1]. Both letters – which later became the major legal documents of the conflict between Leibniz and Newton – were part of the *Commercium Epistolicum*; they were then called *Epistola Prior* and *Epistola Posterior*.

III-A *EPISTOLA PRIOR*, OR LEIBNIZ'S DISCOVERY OF SYMBOLIC FORMS WITHOUT A SUBSTANCE (JUNE 1676)

Like most of his contemporaries, it was through Descartes' *Géométrie* that Leibniz discovered, in Paris (around 1674) the (quite new at that time) symbolical writing of mathematics[2]. In a second step, he met a lot of essential syntactical and semantic problems, raised by the *Epistola Prior* of 1676, a letter that Newton sent him via Oldenburg, secretary of the Royal Society[3]. One must underline the importance of the *Epistola Prior* for Leibniz: at first, in his youth, it was, according to his own terms, a major component of the constitution of his "symbolic thought". Apparently, the *Epistola Prior* was closely dependent on a specific mathematical symbolic form (namely, an exponential that we will call 'Newtonian'; see *infra*). In fact, it soon became for Leibniz a source of questions of a higher philosophical order, such as the respective places of necessity and contingency within the definitions of mathematical objects, before being transformed, at the end of his life, into an element of the trial for plagiarism instituted against him by Newton.

---

1   This chapter is largely based on [Serfati 2011c], 268–286.
2   See [Serfati 2005] (second part), *Symbolique et Invention*.
3   See [GBM], 179–203. On June 13th, 1676, Newton sent his letter to Oldenburg, who forwarded it to Leibniz on July 26th.

Fractional exponents

In the history of mathematics, the importance of the *Epistola Prior* (June 1676) usually stems from the 'binomial infinite formula' proposed by Newton, a result now considered as one of the first power series developments of a usual function[4]; in the late seventeenth century, it was mainly viewed by its author, in a vision now somewhat surprising, as a convenient way 'to extract square roots', a hitherto difficult or impossible operation[5]. However, our project is not adding one more analysis to the many others[6] of the contents of a widely studied formula in the history of science, but to examine the symbolic issues it raises, and how Leibniz discovered them. After a brief introduction, Newton abruptly delivers to Leibniz a general result as follows[7]:

$$\overline{P+Q}^{\frac{m}{n}} = P^{\frac{m}{n}} + \frac{m}{n}AQ + \frac{m-n}{2n}BQ + \frac{m-2n}{3n}CQ + \frac{m-3n}{4n}DQ = \text{etc.}$$

A formula – a 'canon', a term of the seventeenth century that we will treat later – that is immediately followed by an interpretative comment we have translated as follows:

> (...) Where P+PQ means a quantity of which one seeks a certain root or a certain dimension or the root of the dimension. P is the first term of this quantity, Q the remaining terms divided by the first and $\frac{m}{n}$ the numerical index of the dimension of this same P+PQ, that is to say, or the whole dimension itself, or the broken dimension (as it is called), either positive or negative. For, as Analysts instead of aa, aaa, etc. used to write a^2, a^3, and similarly for $\sqrt{a}$, $\sqrt{a^3}$, $\sqrt{c \cdot a^5}$ [8], I write $a^{\frac{1}{2}}$, $a^{\frac{3}{2}}$, $a^{\frac{5}{3}}$, and for $\frac{1}{a}$, $\frac{1}{aa}$, $\frac{1}{a^3}$, I write a^{-1}, a^{-2}, a^{-3}.

In first analysis, the 'canon' thus appears as an equality between, on the one hand, the substance of the left form $\overline{P+PQ}^{\frac{m}{n}}$, and on the other hand, that of the right one, which was then named, in Newtonian terms, an 'infinite series'. Let us repeat, we do not examine here the significance of this equality, but only consider the interpretation of the left form, that the Newtonian introductory commentary above aimed to define. First, let us briefly underline a strictly combinatorial point, obviously epistemologically crucial, but that we will not detail here: by using $\frac{m}{n}$ in the exponent, Newton's formula offered to Leibniz (what I call) a *letteralization* in the place of

4   In modern terms, $(1+x)^\alpha = \sum_{n \in \mathbb{N}} \frac{\alpha(\alpha-1)\ldots(\alpha-(n-1))}{n!} x^n$ the radius of convergence is equal
     to 1 if α is not a natural number.
5   For example, calculating $\sqrt[3]{1+x}$ using the series expansion above, where α = 1/3. I developed
     Newton's Example 1 in [Serfati 2005].
6   See for example [Pensivy 1986]. This thesis contains an extensive bibliography.
7   [GBM],180. I have tried to respect the original typography of Gerhardt's edition. "Dimension"
     is for Newton what we now call an integer power, and "broken dimension" is a rational power.
8    $\sqrt{c \cdot a^5}$ is equal to $\sqrt[3]{a^5}$.

the exponent (i. e. a literal exponent, sign of indeterminacy)[9]. Ignored by Descartes, the creator of the exponential, this indeterminacy of the exponent was indeed a significant conceptual revolution.

Symbolism without interpretation?

The central point here, for us, is different: in Leibniz' time, the symbolic form $\overline{P+PQ}^{\frac{m}{n}}$ was deprived of an object, because deprived of a substance. Each of its inner symbolic forms could certainly be interpreted, and first

P + PQ which, *rightly completed*[10] by round brackets and a 'point', leads to (P+(P. Q)), simply interpreted if P and Q are signs of arbitrary numbers. Similarly, if m and n are signs of integers, $\frac{m}{n}$ is an elementary symbolic form (that is to say, it contains only one operation sign, which we call here, following Bourbaki, an *assembler*), interpreted as a rational number (a *broken number* in Newtonian terms).

Rightly completed, the symbolic form is thus written $\overline{P+(P\cdot Q)}^{\frac{m}{n}}$, the aggregation being indicated, as the case may be, by brackets or by a line over the signs (*vinculum*)[11]. The only question that remains, but it is a major one, is the meaningless character of the highest-level assembler, namely 'the exponential copula'[12].

So, the procedure that might be partly conceived could not however ultimately take place and produce a value. Nothing in Leibniz' previous experience at that time, no more than in the Cartesian definition of the exponential – obviously, the only one Leibniz knew – could let him foresee what meanings Newton might provide to symbolic forms such as $3^{\frac{1}{2}}$, or $(x+3)^{\frac{1}{2}}$, or $5^{\frac{2}{3}}$, or even $\left(\sqrt{2}\right)^{-\frac{6}{7}}$, the interpretable character of which should, however, have resulted from the formula. Any attempt of a rhetorical translation in the Cartesian mode led to a nonsense: if the procedure of the 'form' 3^5 may, indeed, be described as 'multiply the number of sign 3 five times by itself', what could really mean $3^{\frac{1}{2}}$: 'multiply this number half a time by itself'? To Leibniz and to the geometers of his time, for whom these symbolic forms were meaningless, the following comment of Newton brought clearly a

---

9   See [Serfati 1999].

10   We say that an assemblage of signs is *rightly completed* when it is provided with all possible punctuation marks. I studied the interpretation of this practice and this concept in [Serfati 2005].

11   On aggregation of signs, see [Serfati 2005], chap. 5, *Ambigüité de l'ordre et signes délimitants*.

12   Indeed, any exponential form, such as a^z obviously contains an operation sign, the exponential copula, identified all together by the *absence of sign* and the up-position of the exponent; from now I shall call it the 'exponential copula'. In [Serfati 2005], one can find a general definition of levels of the assemblers inside a symbolic form. With respect to our example, we will only specify their values on a lower line:

$$(P+ (P.\ Q)) * (m/n)$$
$$2 \quad 1 \quad 3 \quad 1$$

value, according to a procedure that we reformulate here in a modern exhaustive rhetorical form[13]:

> "The procedure to define the symbolic form $a^{\frac{m}{n}}$ is the following: take the nth root of the positive number a and objectify the result. Raise this result to the mth power"[14].

So, the value of $a^{\frac{1}{2}}$ is that of $\sqrt{a}$; that of $a^{-\frac{2}{3}}$ is equal to $\dfrac{1}{\left(\sqrt[3]{a}\right)^2}$.

This definition was immediately accepted by all the mathematical community, spread through Europe, and was treated by Leibniz in all the public founding texts of the differential calculus, (for example in the *Nova Methodus* (…) of 1684)[15]. And what I from now on call the *Newtonian exponential* is, even today, a universally accepted definition.

On the consistency (?) of Newton's exponential

The questions remained obviously open, of the origin and (especially!) of the relevance of this sophisticated symbolism of fractional exponents, which must have appeared so strange to its first readers, such as Leibniz. In the *Epistola Prior*, Newton gives no explanation, no more than he demonstrates the binomial theorem itself. In various texts, I detailed the epistemological principle at work, which I called 'permanence-ramification'; it is a mechanism that is both very common and very general, both syntactic and semantic (see below). I shall examine here only a single theoretical point, by underlining simply that Leibniz, in the marginal notes on the letter reported by Gerhardt, settles, about this extension, the real 'good question', which I summarize here in a simple example: by following the definition of Newton, what should one say about the comparative values of a^3 and of $a^{\frac{6}{2}}$? With his usual perspicacity, Leibniz indeed raised this immediately, in one of the handwritten notes – printed by Gerhardt – with which he covered the margins of Newton's letter[16].

We detail this problem a little here. Given a rational number, there are infinity many couples of integers whose quotient is equal to it. Let us repeat, the question is then this one: if one has $\dfrac{p}{q} = \dfrac{s}{t} = r$, what about the value of a^r ?

---

13   In the case where a is the sign of any positive number, and m and n are both signs of positive integers.

14   The procedure Newton described is however potentially ambiguous in that it does not specify an order to the execution: one might as well start by raising to the power m, objectify the result, then take the nth root of it. It turns out that the two procedures lead to the same value, therefore there is no ambiguity; this fact being neither demonstrated, nor even suggested by Newton. If one of the two integers, for example the one of sign m, is negative, then by setting $m = -p$, where p is the sign of a positive integer, we start by replacing m by p in the previous procedure, which is legitimate; one then takes the *reciprocal* of the last result.

15   *Acta Eruditorum*, Leipzig, 1684 = GM V, 220.

16   [GBM], p. 180.

We have, indeed, on the one hand $a^r = \left(\sqrt[q]{a}\right)^p$, on the other hand $a^r = \left(\sqrt[t]{a}\right)^s$.
So, the geometer would receive naturally two substances for the same symbolic form: they could be different, which would be contradictory. It can be simply shown that is not the case here[17].

It will be observed at the same time that the question is actually substantially equivalent to another: to an integer with sign m is associated its Cartesian exponential with sign $a^m \cdot$ On the other hand, any integer m is a rational number, and it is true of an infinity of manners, since for every nonzero integer k, $m = \dfrac{m}{1} = \dfrac{k \cdot m}{k}$. We will then show, exactly what had attracted the attention of Leibniz, that the value of a^m (in the Cartesian sense) and that of $a^{\frac{k \cdot m}{k}}$ (the Newtonian sense) are the same for all k, for example, that the values of a^m and $a^{\frac{k \cdot m}{k}}$ above are the same: this fundamental issue is the consistency of Newtonian exponential. Indeed, if we set $\alpha = a^{\frac{k \cdot m}{k}} = \sqrt[m]{a^{k \cdot m}}$, then $\alpha^k = a^{k \cdot m}$, and therefore $\alpha = a^m$.

'Permanence-Ramification'. A scheme

The universal validity of the previous proposition is obviously crucial for the consistency of the definition, which Leibniz had actually seen quite well. We must here emphasize the contingency of the definition of Newton: the substance of $a^{\frac{p}{q}}$ could not-be, or be-different from the one Newton proposed. However, and as soon as Newton displayed a definition, whatever it was, it was necessary that the values of a^3 and $a^{\frac{6}{2}}$ were in all cases the same. From 1676 thus (and with the strength of its simplicity), the Newtonian example so addressed to Leibniz a lesson of concrete philosophy which articulates a real distinction between necessity and contingency. In a specific article ([Serfati 2011c])[18], I had repositioned this 17th-century example in a wider, standard issue, of creations of objects, including in contemporary mathematics, *via* the procedure of 'permanence' (of the symbolism) versus 'ramification' (of the meanings) – this leans on the (contingent) choice of certain mathematical formulae (which I referred to in the text as '*canons électifs*').

---

17 By setting $b = \left(\sqrt[q]{a}\right)^p$ et $c = \left(\sqrt[t]{a}\right)^s$, one gets $b^q = a^p$ and $c^t = a^s$, so that $b^{q \cdot s} = (b^q)^s = (a^p)^s$
   $= (a^s)^p = (c^t)^p = c^{t \cdot p}$. One has $q \cdot s = t \cdot p$. Thus $b^{q \cdot s} = c^{q \cdot s}$ and $b = c$.

18 'Analogies et "prolongements". Permanence des formes symboliques et constitution d'objets mathématiques.'

III-B THE *EPISTOLA POSTERIOR* (OCTOBER 1676)

Irrational Exponents

Leibniz was certainly impressed by the *Epistola Prior*, as evidenced by his some-what embarrassed response of August 27th[19]. Newton replied in turn on October 24th, 1676. In its last paragraphs[20], this letter, now called *Epistola Posterior*, contained, for Leibniz, a new Newtonian expansion of the exponential, drafted in these terms:

Communicatio Resolutionis Affectarum per Methodum Leibnitii pergrata erit, juxta et Ex-plicatio Quomodo se gerat, ubi Indices potestatum sunt Fractiones, ut in hac Aeaquatione

$$20 + x^{\frac{3}{7}} - x^{\frac{6}{5}} y^{\frac{2}{3}} - y^{\frac{7}{11}} = 0 \quad \text{aut Surdae Quantitates, ut in hac } \overline{x^{\sqrt{2}} + x^{\sqrt{7}}}^{\sqrt{(3)^{\frac{2}{3}}}} = y \quad \text{ubi } \sqrt{2} \text{ et } \sqrt{7}$$

non designant Coefficientes ipsius x, sed indices Potestatum seu Dignitatum ejus $\sqrt{(3)\frac{2}{3}}$ indi-cem Dignitatis binomii $x^{\sqrt{2}} + x^{\sqrt{7}}$.

And Newton added: "*Res, credo, mea methodo patet; aliter descripsissem*". He so proposed to Leibniz a new meaning for the symbolic form a^r, via a new extension of the object-field of the letter r, the substance now permitted to the exponent being the 'surd' numbers (for example a quadratic irrational like $\sqrt{7}$). The example of Newton, which has absolutely no interest other than rhetorical and 'educational' – is however particularly suggestive: it indeed contains a superposition of two such exponents, one within a elementary symbolic form (the binomial), the other in the exponent of the binomial. A Newtonian writing that certainly did not fail once again to surprise Leibniz! As to the meaning of his new exponential, this time Newton furnishes none. The last lines are cryptic: the matter, Newton says, follows ob-viously from his method and he will give the description somewhere else!

Letters in Exponents

At the end of 1676, the situation of Leibniz was as follows: having just discovered the mathematical rhetorical writing of his youth, Greek and scholastic, he had met two years earlier, in Descartes' *Géométrie*, the new symbolic writing, in which the Cartesian exponent was of a use quite new at the time. Let us remember that the exponent, introduced by Descartes in the *Regulae*, had then been popularized by the text of *La Géométrie*. Let us also remember how important was, in the Cartesian conception, the fact that in the place of the exponent was settled a simple integer, such as '$2a^3$' from Rule XVI of the *Regulae*[21]. Leibniz had to have become familiar

---

19   [GBM], 193–203.
20   *Ibid.*, 225. I tried to reproduce the typographical arrangement of the Gerhardt edition.
21   In this text indeed appeared, under Descartes' pen, the first effective exponent in history. See *Regulae ad Directionem Ingenii*, AT, X, 360–469, here 455. Also see the French translation [*Regulae*]. About exponents in Descartes, see [Serfati 1998] and [Serfati 2008c].

quickly and deeply with the new symbolic system, extremely distant, however, from the considerations of his youth. Yet, two years after his reading of the *Géométrie*, and in less than four months (July–October 1676), he had, this time because of Newton, to integrate within his conceptions two still new exponential forms, obtained by two successive field extensions (broken and surds), grounded on the same Cartesian symbolic form, moreover *letteralized* by Newton. That the first version of the Newtonian exponential, the *broken* one, was correctly and completely defined in *Epistola Prior*, whereas the second, the *surd* one, was in no way explained in *Epistola Posterior*, did not constitute – as one will see – a concern for the Hanoverian mathematician, who deduced this double conclusion: on the one hand, the Cartesian exponent was certainly not fixed, contrary to what the strength of a recent tradition had tended to persuade. On the other hand, Leibniz concluded that all questions of meaning were actually subordinate, when compared with the power of symbolic realities (namely, the primacy of the exponential form). With that, capturing the essence of the Newtonian method, Leibniz dedicated himself from then on to building a still new exponential 'form'

$$a^z \text{ or } x^z$$

whose importance would even exceed that of Newton's (at least he hoped so). One can see this approach at work, right from the letter to Tschirnhaus of May 1678.

III-C DESCARTES, LEIBNIZ
AND THE IDEALIZED IMAGE OF THE EXPONENTIAL

Gradus indefinitus

The question then was this one: what could be, at that time, and for him, the value of a symbolic form where, in the place of the exponent, there came the sign of an *indeterminate* number? Such an exponential, we shall say 'Leibnizian', constituted for him one aspect of what he called *transcendence*, in the mathematical sense. This point will be developed in detail below.

Let us return at first to the letter to Tschirnhaus already mentioned above, dated May 1678[22], a year and a half after Leibniz' departure from Paris. Leibniz proposed for the first time an *indefinitus gradus* with the symbolic form[23]:

$$x^v + \overline{ya}^x + \text{etc. aequ. } 0$$

From then on omnipresent in the correspondence of Leibniz, this exponential will be anecdotal by no means, but on the contrary, by a genuine idealization of the exponential symbolism, one of the major models of what he will call *transcendence*[24].

22   [GBM], 372–382.
23   Fac-simile of [GBM], 377.
24   See [Breger 1986]. 122–123.

This symbolic invention had a double historic origin, both in Descartes and in Newton, two scholars of whom Leibniz declares clearly, in diverse texts, that with his exponential, he exceeded them both. First as to Descartes: we measure today insufficiently the progress and the revolution brought by Cartesian exponential – I have described them in diverse texts[25]. Descartes had elaborated it against the inadequacies of the cossic system, and it was included for a while in all the new calculation of the first half of the XVIIth century. In reality, the symbolic texture of the *Geometrie* had appeared to its contemporaries as clearly revolutionary, because of the Cartesian exponential. An exponential, which ordered and spontaneously organized into a hierarchy what we now call the polynomials, and certain plane curves (contrary to what Hofmann writes. See the *Introduction* above). However, by producing his two exponentials, broken and surd, in both *Epistolae*, Newton had then far exceeded Descartes in Leibniz' eyes. These extensions probably convinced him both that the essence of any calculation stood in the exponential, and that the structure of the latter was not definitively fixed. By his new symbolism a^x, he could thus hope to exceed (that is, *transcend*, literally) both Descartes and Newton. A symbolism that, as far as I know, he was more or less the only one, at that time, to write and to claim continuously (later Jean Bernoulli will follow suit)[26].

Descartes and Newton transcended by Leibniz

Hence Leibniz considered at first that Descartes was correct against the 'Ancients' because he 'welcomed exponents in his geometry', but he also considered that the latter had not gone far enough with this exponential point. Here is Leibniz's argument, in a very clear text (1684)[27]:

> "The Ancients did not want to use the lines of a higher degree nor the solutions that stemmed from them (...) Descartes criticizes this and accepts in geometry all the curves whose nature is Algebraic in some equation, or which can be expressed by a fixed degree. This is admittedly correct, but he will commit a mistake in it which is of no less importance than that of the Ancients (...) because he excludes all the other infinite curves from geometry, even those that

---

25 See in particular [Serfati 2005], Chapter IX: *Puissances, de Descartes à Leibniz. La représentation des concepts composés* (p. 221), Chapter X: *Le système symbolique de Descartes* (p. 235), and Chapter XI: *L'exponentielle après Descartes* (p. 251).
26 About the invention of the signs by Leibniz, cf. M Dascal: For Leibniz, the signs are invented not only for communication, but are created "to help in the invention and guide the judgment". He thus writes: "De même que ce n'est pas dans les fonctions cognitives "inférieures" (évocative, abréviative, mnémonique) que réside le grand avantage d'une notation mathématique adéquate, ce n'est pas non plus dans sa fonction communicative. Leibniz, bien sûr n'oublie pas que les signes mathématiques servent aussi à la communication. Mais il envisage cette fonction toujours comme secondaire par rapport aux fonctions cognitives 'supérieures' (jugement, raisonnement, invention). Des signes qui, comme ceux des opposants de l'algèbre de Viète ou ceux des astrologues, alchimistes et tant d'autres, ne sont inventés que pour l'abréviation, ou la communication, ou pour garder des secrets, sont incomparablement inférieurs à ceux qui sont créés pour aider à l'invention et guider le jugement" ([Dascal 1978], 217–218).
27 *De dimensionibus figurarum inveniendis*=GM V, 124, (1684).

can be described with precision, because, doubtless he was unable to turn them into equations in order to apply his rules to them; but it is really necessary to know that also even those, such as the Cycloid or the Logarithmic, and others of the same kind, of which we make the greatest use, that can be expressed by calculation, and even by finite equations, but obviously not by algebraic equations, that is to say of finite degree; but of indefinite degree, that is to say transcendent; thus they can be submitted to calculation, just in the same way as the other ones; it is possible that this calculation is of another nature than the one which is commonly used."

Descartes being so widely inadequate, Leibniz then declared that he prefered Newton even if, for him, the latter did not go quite far enough either. As the *Epistola Posterior* had shown abundantly, Newton indeed welcomed irrationals. So Leibniz will write to Wallis in 1697[28]

"(...) of very perceptive Newton who called Geometrical irrational those that Descartes does not welcome in its geometry; but these, I distinguish them from the transcendent, in the same way as kind from species."

But surpassing Newton was also part of Leibniz' process and symbolic strategy! It was indeed in this precise point of the *letteralized* exponentials that he placed the originality and the strength of his creation of these archetypes of transcendence. These expressions really came to embody, for Leibniz, the essence of his 'new calculation', that of the 'transcendent'. Then all this will be about transcendent expressions (*i. e.*, functions) with *gradus indefinitus*. Thus, we can somewhat understand Leibniz's unbelievable (and sometimes unreasonable) attachment to those exponentials, a point we will re-examine thoroughly further in the book[29].

CONCLUSION OF THE FIRST PART: THE TRANSCENDENT IDENTIFIED WITH THE NON-CARTESIAN FIELD

I gather together here some conclusions for this first part. As we saw in detail, the constitution of the term and of the concept(s) of transcendence for Leibniz was co-extensive with his progressive awareness of the active existence of a non-Cartesian domain in mathematics, a vast and essential field, although with ill-defined borders. I summarize five important stages in the elaboration of Leibniz's conceptions:

1) His development, in 1673, of the proof for the arithmetical quadrature of the circle (following Mercator) and the discovery of an expression for the '*numerus impossibilis*': of course, the total sum of the series represents the required value, but in a very specific way – it was not a Cartesian one (see above, Chap. II-A), a fact that Leibniz was immediately convinced of. He so discovered an initial failure of

---

28 Leibniz to Wallis (1697) = GM IV, 28.
29 Leibniz would have wished that all transcendent functions fall into this privileged case, but he realized gradually that's not the case. So in later texts, he changes the structure of his definitions. In a letter to Johann Bernoulli (GM III, 319) he calls 'percurrent expressions' (*percurrentes*) the most general non-algebraic expressions: "They are to me like the kind," he says, and "transcendent", the exponentials such as x^y, of which they are "the most finished species". See below in Chap. V-E): Percurrents and transcendents. Leibniz and Johann Bernoulli.

Descartes as regards the numerical domain (see below, Chap. VII, *Leibniz and the Transcendent Numbers*).

2) The controversy between Gregory and Huygens, following the publication of *Vera Circuli* in 1667, was known by Leibniz in 1673 (see above, Chap. II-B). Gregory approached very closely the problem of numerical transcendence (in the modern sense). However, we will have to wait for Lambert as regards a precise definition of the transcendent numbers (see below, Chap. XIII-C). What is important for us here is that Leibniz did not doubt the result, namely the non-algebraicity of π. Accordingly, Leibniz strengthened his beliefs about the inadequacy of the Cartesian numerical conceptions.

3) The first draft of the letter to Oldenburg, March 30th, 1675. Here, Leibniz uses 'transcendent' to mean 'what is not analytical', that is, in this first instance, that which is not subject to a Cartesian symbolism. The issues regarding quadratures, here, motivate the use of the word. Both he and Mercator, however, had managed to carry out fundamental quadratures – those of the hyperbola and of the circle. However, their results did not fit into the Cartesian 'analytical' framework.

4) Both *Epistolae* dated June and October 1676, sent by Newton, via Oldenburg. Leibniz was fully interested by such a focus on some large developments in power series as a major mathematical tool. These new aspects undermined some of his convictions, because at that time he was hardly prepared for them. In retrospective, these developments made him consider the quadratures of the hyperbola and of the circle as particular developments in power series[30]. This intervention of the infinite, which is unavoidable here, was entirely outside of the field of Descartes' conceptions! However, and above all, Leibniz learned from these two *Epistolae* that the Cartesian exponential, which had contributed so much to mathematics – it had put a definite end to the Cossic system – was not symbolically rigid and could be subject to major extensions. There were two successive extensions of the exponential which Leibniz was greatly interested in. This was a particularly important lesson for Leibniz, since it was a part of the symbolic field.

5) The letter to Tschirnaus, dated May, 1678, following the two *Epistolae*.

Accordingly, during those early days of the transcendence for Leibniz, the aspect of letteralized exponentials – the sign of which was a^x – became a major part of his mathematical thought – indeed, this fact is somewhat obscured today. This idealizing symbolic conception, at the same time as it clearly denoted the high consideration granted by Leibniz to Descartes' design (with the creation of the primitive exponential), radically parted from it. In fact, with his letteralized exponentials, Leibniz aimed at embodying *in situ* Descartes' invention – even exceeding it. We can now better understand the reasons why Leibniz would never cease in repeating how his resolutely non-Cartesian exponentials – which were no more Newtonian –

---

30   This is a theme that Leibniz will never cease in resuming afterwards. For example, in a series of epistolary exchanges with John Bernoulli between 1696 and 1698, both correspondents were interested in searching for differential equations verified by the sums of some power series (with an instantiation to the sums of some numerical series) (see A III. 7, N. 46, N. 47, N. 54 and N. 88). This idea is fully in line with the method of the arithmetical quadrature of the circle.

represented a peak of the notion of transcendence (see Chap. IV-D, *The exponential symbolisms*).

Thus, during these early years (1673–1678), Leibniz gradually discovered the pregnant existence of a new mathematical universe which was in no way ruled by Descartes' conceptions, whatever their very considerable merits might have been. In Leibniz's thought, the concept of transcendence was then gradually and necessarily identified with that of the non-Cartesian field. However, such a founding definition is fully negative, that is to say, indirect. It was, therefore, natural to try to capture directly, in the best possible way, its domain of existence. As such, the project to make an – even partial – inventory of the (unknown) non-Cartesian field was equally natural. As we see below, this is what Leibniz would try to do in a second step.

At the end of this first period, we can thus note some main emerging lines, both in Leibniz's forthcoming mathematical research and in the evolution of his conceptions.

SECOND PART
THE SEARCH FOR AN INVENTORY

CHAPTER IV
FROM INFINITELY SMALL ELEMENTS
TO THE EXPONENTIAL UTOPIA

IV-A TSCHIRNHAUS AND THE INVENTORIES OF 1679–1684

After this initial phase, Leibniz paused for a moment, trying to *delimit* the notion of transcendence he had just brought to light and then named. Indeed, the real issue was to delimit and not to define. It was, for Leibniz, confronted with the immense and unknown territory (also negatively defined) of non-Cartesian knowledge, an attempt to inventory certain regions. In other words, Leibniz had to give positive content (and the most exhaustive such content possible) to his own terminology. From this perspective, Tschirnhaus was, for a few years, his privileged interlocutor.

I shall examine first a letter, dated 1679, from Leibniz to Tschirnhaus[1]. In it, Leibniz is very clear about his attempt to inventory transcendence according to three aspects[2]. The letter thus marks an epistemologically important step in the process of the domestication of the unknown. One of the difficulties of this work will arise from the fact that the area of the inventory is naturally very diverse, involving transcendent objects, numbers, curves, and expressions. Leibniz states:

> Cæterum tres habeo calculos transcendentibus etiam applicabiles, unum per differentias seu quantitates infinite parvas, alterum per series infinitas, tertium per exponentes irrationales, ex quibus novissimus habet aliquid prae cæteris, per ipsum enim solum demonstrare possum impossibilitatem quadraturae specialis ex. gr. circuli totius; per duos vero priores tantum invenire possum aut impossibilem demonstrare quadraturam generalem algebraicam alicujus figurae (GM IV, 480).

As such, Leibniz proposes an inventory of transcendence in three parts:

1°) The use of infinitesimals (transcendent objects).
2°) The infinite series (transcendent expressions).
3°) The use of fractional exponentials – they will later become letteralized (transcendent expressions).

Five years later (September 1684), in another letter to Tschirnhaus[3], Leibniz resumed, almost with the same words, this tripartite classification:

> Generaliter autem Calculus Transcendens mihi est triplex. Est enim adhibenda æquatio quoad numerum terminorum vel infinita vel finita; si infinita, tunc proveniunt series infinitæ, quas jam et alii ante me adhibuere, etsi in illis nova quædam magni momenti detexerim. Si æquatio numerum terminorum habeat finitum, tunc rursus vel adhibet quantitates infinitas infiniteve par-

---

1   GM IV, 477–483 = GBM, 401–406.
2   This point is noted in [Breger 1986], 125.
3   GM IV, 507= GBM, 457–462.

vas (tangentium tamen ope in ordinariis repræsentabiles), quod speciatim facit Calculus meus differentialis, vel adhibet quantitates ordinarias, sed tunc necesse est ut incognitæ ingrediantur exponentem, et hanc ultimam expressionem omnium Transcendentium censeo perfectissimam, hanc enim ubi semel nacti sumus, finitum est problema (GBM, 459)

This ternary division was central to Leibniz. We examine in more detail in this chapter some of these Leibnizian procedures. Naturally, none is Cartesian! Yet it should be noted that this tripartition does not address the question of transcendent curves, equally fundamental at the time (and for Leibniz himself). However, Leibniz will also develop this topic in many other texts! We return to this point later, in Chapter V (*Geometry and transcendence; the curves*), and shall make here two terminal remarks about the second and third components of the above inventory.

First, the use of power series (the second component). To characterize transcendence in the present case, one must demonstrate that one cannot use polynomials (which contain a finite number of terms) and that the use of an infinite sequence or series is thus unavoidable. However, this is precisely what Leibniz had already been interested in, through some earlier texts associated with his arithmetic quadrature, such as *Praefatio Opusculi De Quadratura Circuli Arithmetica*, dated 1676 (GM V, 93–98): one cannot have, says Leibniz in this text, a universal equation (i.e., of fixed finite degree) between x and $\sin x$, for example[4]. This is actually to demonstrate that $\sin x$ is a transcendent function in the modern sense. On the other hand, as regards the third component of the inventory above, it should be emphasized, in the letter *supra* dated 1684, as to the strong opinion of Leibniz according to which letteralized exponentials represent "the most perfect expression of all the Transcendent."

The following two sections, IV-B) and IV-C), developed the first and third components of the inventory above.

IV-B DE BEAUNE, DESCARTES, LEIBNIZ, AND THE INFINITELY SMALL ELEMENTS

When a curve is no longer considered as "a set of points" (in modern terminology)

The second of the problems of geometry that Florimond De Beaune submitted to Descartes in 1638, shortly after the publication of the *Géométrie* (see letter from Descartes to Beaune, dated February 20th, 1639 (AT II, 514–517), is of central epistemological importance in the history of mathematical ideas. This type of problem was completely new at the epistemological level. For the first time in history, a curve was conceived – indeed, here – not as the *set of all its points* but – we can better say, in modern terms – as *the envelope of its tangents*. From then onwards, for Descartes, such a curve "is not given yet," even if De Beaune succeeds in calculat-

---

4   "Sed relationem arcus ad sinum in universum aequatione certae dimensionis explicari impossibile est," (GM V, 97). This point is also raised by Breger (page 125).

ing it! Accordingly, he writes to De Beaune that "[Archimedes] examined a given line, instead you determine space contained in another one, which is not given yet." Milhaud comments here, somewhat pertinently[5]:

> This is the first historical example of curved lines which are not defined by a characteristic property of their points, or, as says Descartes, which are not given yet, on which, however, we solve such or such problem by appealing to a property of the tangents to such (supposedly determined) curves.

This problem deals with the determination of one (or more) plane curve(s) by a condition on the tangent at the current point. In modern terms:

> Given an axis AY and a point C (AC = b) on the perpendicular on A to the axis, one has to find a curve such that the tangent to an arbitrary point, say X, is parallel to the straight line joining C to the intersection of the axis AY and of the line containing X forming an angle of 45° with the axis[6].

This problem is typical of what was later called the 'inverse method of tangents'[7], a procedure which largely occupied mathematicians for the second half of the seventeenth century. It leads, in modern terms, to a differential equation, which it is then a question of solving (if one can)[8]. In this particular case, it is the linear equation of the first-order[9]:

$$\frac{dy}{dx} = \frac{x-y}{b} \quad (E)[10]$$

On the other hand, and as Breger rightly stresses (p. 129), De Beaune's problem has always been a point of fundamental methodological difference between Descartes and Leibniz. It is indeed clear that, for Leibniz, the rhetorical value of this problem remained important, both as to the status of his infinitesimals as well as the superiority of his mathematical thought to that of Descartes.

Solving the differential equation (E) typically leads, in modern terms, to a family of curves with one parameter (below denoted μ). We sketch the method by using the conceptions given by Leibniz. To solve (E), Leibniz makes – basically – the change of the unknown function $y \to s$, where s is defined by:

$$s(x) = x - y(x) - b$$

---

5   [Milhaud 1921], 170.
6   See AT II, 520, and also [Milhaud 1921], 170–171. In his letter to Beaune dated February, 1639, Descartes does not remind us of the statement of the problem, which he will clarify, however, in a later letter to Mersenne (in June, 1645) (AT IV, 229). Regardless, so formulated, the statement of the problem contains an ambiguity of sign which is due to two possible interpretations of the expression "forming an angle of 45° with the axis."
7   In 1676, Leibnz also speaks of "Méthode inverse des touchantes" (GBM, 201).
8   See *infra* Chap. VI, *Quadratures, Differential Equation, and Inverse Method of Tangents.*
9   A first-order differential equation is said to be linear if it can be written as:
$$a(x)\,y'(x) + b(x)\,y(x) = c(x).$$
(Equation (E) is thus obtained here by setting $a(x) = b$; $c(x) = x$ and $b(x) = 1$). All these equations can be solved (up to the evaluation of some integral) by a standard method, namely the "variation of parameters."
10   AY was selected for the ordinates' axis. See AT. II, 520. See also Milhaud, 171.

One gets $y(x) = x - b - s(x)$; hence, $\dfrac{dy}{dx} = 1 - \dfrac{ds}{dx}$

On the other hand, (E) can be written as:

$$\frac{dy}{dx} = \frac{x-y}{b} = \frac{b+s(x)}{b} = 1 + \frac{s(x)}{b}.$$

Thus, $\dfrac{ds(x)}{s(x)} = -\dfrac{dx}{b}$, and this is a 'logarithmic' differential equation, that is to say, a truly canonical equation in the eyes of Leibniz[11]. Its solutions are $\ln\left|\dfrac{s(x)}{K}\right| = -\dfrac{x}{b}$.

Finally, $s(x) = Ke^{-\frac{x}{b}}$, or else:

$$y(x) = x - b + \mu e^{-\frac{x}{b}}$$

Descartes and De Beaune's problem

The equation thus admits among its solutions exponential curves (which Leibniz calls "logarithmic"). Descartes had obviously not used a differential equation (!), but heuristic kinematic considerations, somewhat risky: he explains that the solution is obtained by the composition of two movements, among which one is uniform and the other is such as its speed increases in proportion to the distance covered, two movements that Descartes, according to his constant position[12], considers "incommensurable." He does not mention logarithms or exponentials, and he rejects the resulting solution in the "mechanical" curves (see *infra* Chap. XV, *Algebraic curves, transcendent curves (modern sense)*).

One is actually a little surprised that Descartes was interested in this issue, which is clearly outside his field as regards the accuracy of knowledge and thought. Very probably he did so because he wanted to respond favourably to Florimond De Beaune, a staunch Cartesian, and one of his firmest supporters. Descartes concludes:

> But I believe that these two movements are so incommensurable, that they cannot be exactly adjusted the one by the other one; and therefore that this line belongs to those that I rejected of my Géométrie, as being only Mechanical; that is why I wonder no more of the fact that I was not able to find it by the other means that I had used because it only applies to the Geometrical lines (AT II, 517, lines 9–16).

---

11 See *infra* Chap. IV-D5. Leibniz finds $\dfrac{ds}{s} = \dfrac{x}{b}$; this doubtless corresponds to the other interpretation of the statement. Nonetheless, the structures of the final results (sign +, or sign -) are nearby.

12 See *Géométrie* (AT VI, 412).

Leibniz and De Beaune's problem

Leibniz evokes, with Oldenbourg, the problem posed by De Beaune in August 1676 (GBM, 201–203) in order to criticize and mock the solution offered by Descartes and, at the same time, to highlight the advantages of his differential calculus. In a supplement to the letter, written in French:

> Mr. Descartes speaks with a little too much presumption of posterity; he said page 449, letter 77, that his general rule to solve all the sursolid problems was, without comparison, the most difficult to be found among all the things that were invented so far in Geometry, and which will still be perhaps in several centuries, unless I myself would take the time to look for other ones (as if several centuries were not able to produce a man who could do such a thing, which does not seem to be of the most considerable ones (GBM, 201).

A little farther on, Leibniz mentions "the Problem of the inverse tangent method, that Mons. Des Cartes claims to have solved" (GBM, 201).

Leibniz will return to the problem of De Beaune eight years later, in the *Acta Eruditorum* of 1684, at the very end of the *Nova Methodus* (GM V, 226), focusing very much on the efficiency of his own infinitesimal method. He details first of all, in the article, his algorithm for the calculation of his infinitesimals and some of their main applications. Next, he makes clear that, for him, Descartes had not solved Beaune's problem, which he himself had succeeded in doing with his infinitely small quantities:

> As an appendix, I would add the solution of the problem, which Mr. De Beaune asked of Descartes, that the latter tried to solve in the second volume of his letters, but in vain.

He then gives, again, the statement of the problem of De Beaune, and then solves it using his infinitesimal methods, which are, basically, those that we described above. Incidentally, we can underline the way in which Leibniz convinces himself that the solution which he obtains is a "logarithmic curve:" It is a simple linking between two progressions, one geometrical and the other arithmetical. This method for identifying a logarithm or an exponential – which might seem to us rather surprising today – is customary in Leibniz (see *infra* the construction of the catenary in Chap. V-D).

IV-C ON EXPONENTIAL SYMBOLISMS

We saw in II–1) above, in particular in the letter to Tschirnhaus of 1678, how following both of the *Epistolae* of 1676, Leibniz had convinced himself of the essential importance of letteralized exponentials, whose symbolism:

$$a^x$$

surpassed both Descartes and Newton. He then develops, in different directions (which sometimes may not appear logically related), the idea that he has identified one of the summits of transcendence, and that this is his own creation. In Sections IV-C) (1) to IV-C) (5), which follow in this chapter, I will distinguish three main aspects in Leibniz: his equations with letteralized exponentials, then his exponential representations of curves, and finally the possibilities of quadrature (if any) via exponentials.

IV-C1 ON THE RESOLUTION OF THE EXPONENTIAL EQUATIONS. THE "ADMIRABLE EXAMPLE"

We saw in Chapter III above that, when he received the *Epistola Prior,* Leibniz had briefly been interested – in his marginal notes – in the consistency of the exponential extension $a^{\frac{m}{n}}$. His question was: if $m' = km$ and $n' = kn$, so that $m'/n' = m/n$, what about the definition of $a^{\frac{m}{n}}$? It was one of his first concerns within the register of the exponentials.

Two years after the reception of both *Epistolae* of 1676, the letter of Leibniz to Tschirnhaus of May 1678[13] contains *one of the first exponential equations*, expressed under this still very general form:

$$x^y + \overline{ya^x} + \text{etc. aequ. } 0.$$

At that time (1678), such an equation was not grounded for Leibniz on some geometric interpretation; no more did it arise from any necessity of calculation! It really has one – purely rhetorical – virtue, namely to exceed (i.e., *transcend*) Descartes and Newton, and at the same time place in full sight a simple combinatorial game, the letteralization by which Leibniz – by substituting letters where Descartes had placed numbers – produced new equations and new mathematical objects (see *supra* Chap. III: *The power of symbolism: Exponentials with letters*).

For Tschirnhaus, Leibniz then focused on solving the new equation that he had just proposed, describing what was, at that time, his philosophy of transcendent equations. I feel it is useful to quote, here, an excerpt already given in Chapter I:

> (...) The one who will be able to solve such equations as $x^y + \overline{ya^x} + $ aequ. 0 either by transforming them into ordinary equations, or by demonstrating the impossibility of such a transformation, will be the one who will perfectly be able, either to find all quadratures, or demonstrate their impossibility, because all quadratures can be expressed by such transcendent equations. That's why we must improve this kind of transcendent calculation, through which we shall absolutely have all we seek in this field (GBM, 377)

This statement is somewhat peremptory; Leibniz, of course, does not provide any attempt at proof (we understand today that it was impossible); however, it shows the confidence of the author in the calculation of the letteralized exponentials that he intends to introduce: what we call *infra* IV-C4), the *exponential utopia*.

13   GBM, 372–382.

The admirable example

From 1678 onwards, Leibniz would continually send his correspondents more examples of such equations, often naively so; first of all, this one, which is present everywhere in his correspondences:

$$x^x + x = 30 \ (1)$$

and which he sometimes gives in the equivalent form:

$$x^x - x = 24 \ (2)$$

This equation, I will call – in an (ironic!) tribute to Leibniz – "the admirable example," so as to underline the extent to which the author tries to use it for its rhetorical value. Leibniz emphasizes, each time and for the benefit of his correspondent, that $x = 3$ is a solution. This result is obviously true given the forms (1) and (2) above, and this seems to be sufficient to him!

As earlier as September 8th, 1679, he had written to Huygens on the subject (GM II, 18)[14]. He begins by considering a system of two equations with letteralized exponentials:

$$x^z + z^x = b \text{ and } x^x + z^z = c \ (3)$$

In these equations, he writes: "b and c being given, we ask x and z." It is a system built in a purely formal way, on the one hand by substitutions of symbols, on the other hand by symmetrization, and which Leibniz is obviously unable to solve! In the continuation of his letter, Leibniz gives up system (3) and returns to a single equation (1), which is, he says, "an easier example"; this is (not surprisingly!) the "admirable example:"

> Let us take an easier example, $x^x - x$ is equal to 24, we ask for the value of x and we shall find that it is 3, because $3^3 - 3$ is 27–3, that is, 24. Here is such an equation which is *nullius certi gradus cogniti*, and whose degree even is queried. One could certainly describe lines, the intersection of which could give the solution of these problems, but I ask for a solution, which gives me the value of the unknown. I implore you, Monsieur, to think of it a little, because you see that these are real determined problems, and there must be in the nature a method to solve them.

This example reveals the manner in which Leibniz conceives, at that time (1679), equations with letteralized exponentials, that is to say, equations where the *unknown enters the exponent* (alternatively, an equation "whose degree even is queried"). Leibniz explicitly excluded the possibility of using curves the intersection of which would give the solution x; we know today that it can nevertheless be a good method, and sometimes indeed it is the only one! Thus, Leibniz does not intend, for example, to construct the curve of equation $x \rightarrow x^x$. Huygens answers him (November 22nd, 1679) (GM II, 28), stating that $x = 3$ certainly is a solution in integers, but that there may be others, in fractional numbers. In his reply (GM II, 31), Leibniz simply

---

14   See [Breger 1986], 126.

asserts, without proof: "I can show that this equation is determined, that is to say, it has a finite number of roots"[15].

From the 1680s onwards, Leibniz added more examples of equations with letteralized exponentials. The abundance of the cases, however, only makes it more obvious the very small number of effective resolutions. As we understand it today, Leibniz was only very rarely able to solve the equations that he proposed! We find very numerous instances of this situation. Thus, an exchange with Tschirnhaus of 1681–1682 concerns the equations $x^y + y^x = a^{xy}$ (from Leibniz to Tschirnhaus, GM IV, 485), and $x^y + y^x = a$ (answer of Tschirnhaus, GM IV, 487), which neither of the protagonists was capable of solving. They then decided to differentiate them (See *infra* IV-D)-(5)).

The impossibility of effective resolutions

This impossibility of solving the general exponential equations was doubtless a little irritating and worrying for Leibniz, to the extent that, in 1692 in a very brief article published in the *Journal des Savants* and dedicated to the introduction of transcendence[16] (GM V, 278–279), he is forced, in order to support his argumentation, to propose an example of an *ad hoc* effective resolution. The issue, absolutely trivial, is that of equation $c^x = ab^{x-1}$ (where $c \neq b$), which obviously admits for a unique solution, $x = \dfrac{\log a - \log b}{\log c - \log b}$! As Breger pertinently notes "For this result in the year 1692, he really did not need any publication" ([Breger 1986], 128). Breger also remarks:

> Leibniz was interested in a very intensive way in the first years which he spent in Hanover in the manipulation of exponential equations (p. 126) (…) Reducing transcendent issues to exponential equations was apparently the maximum program of Leibniz. This would have had the substantial advantage of not revealing too significantly the break with Descartes' Mathematics: instead of constant exponents one would have also simply accepted variable ones in the geometry. But Leibniz obtained no usable result in all his hard work on the exponential equations (page 127).

Considering the reduction of all questions of transcendence to exponential equations would be a characteristic approach of Leibniz; it was obviously vain as we understand it today. This granted superiority – both by Leibniz and, later, by Jean Bernoulli – to exponential equations with letters (expected to provide the key to all new mathematical problems) reveals a *symbolic avatar* of Leibniz's inventory process. It carries a permanent trace of the symbolic legacy of Descartes through the exponential (see *supra* Chap. III). Let us note here that there will also be, for Leib-

---

15 The result is nevertheless correct. One can simply demonstrate that $x = 3$ is actually the only solution.

16 *Nouvelles remarques touchant l'analyse des transcendantes, différentes de celles de la géométrie de M. Descartes.*

niz, a *geometric avatar* of his inventory approach, namely the *percurrent* curves (see *infra* Chap. V-E).

IV-C2 EXPONENTIAL EXPRESSIONS AND THE DIALECTICS OF INDETERMINACY – THE STATUS OF THE LETTER

The symbolism of quantities "arbitrary, but however fixed"

In the creation of his transcendent exponentials, Leibniz, as we have seen, had operated, first of all, by substitutions: with letters instead of numbers. But what about the meaning of the letter? I begin, here, a brief but necessary digression upon the epistemological status of the letteralized exponentials for Leibniz. In a number of philosophical and epistemological texts[17], I highlighted what I have called the *dialectics of indeterminacy*, to which every apprentice mathematician, since Vieta, has to submit; I will summarize as briefly as possible this major issue[18].

Since the introduction by Vieta, in 1591[19], of a letter such as '*a*' or '*x*' in the calculation, this letter was supposed to represent an *indeterminate* quantity in the mathematical symbolism, that is to say a quantity which is *at the same time fixed and arbitrary*. However, such a statement is a contradiction in natural language terms. The continuation of any mathematical work equipped with a symbolism of letters can thus be undertaken only by acceptance within the framework of the symbolic language, and not in that of any natural language, of this inaugural agreement: *there exist, in mathematics, arbitrary and fixed entities*. This issue between natural language and symbolic language, to which one must add the acceptance (in the symbolic language) of the aforesaid convention, characterizes precisely what I called 'the dialectics of indeterminacy'.

Through the creation of his **letteralized** exponentials, Leibniz could not avoid being directly confronted with the same problem of indeterminacy. However, one can easily understand that he has not been able, in the late seventeenth century, to identify it as explicitly as I have just done here. Let us recall that the first relevant reflections on the issue will appear only at the beginning of the twentieth century, in Frege and Russell, for example[20]. It was therefore natural that Leibniz would vacillate and oscillate (without deciding) as to the status of the exponent, i.e., of the letter x in the exponential a^x.

17   [Serfati 2010b], [Serfati 2006b], [Serfati 2005], [Serfati 1999].
18   See [Serfati 2005], Chapter VII. *Viète et la dialectique de l'indéterminé. La représentation du "donné"*, pp. 145–197.
19   See [Viète 1591].
20   See *La contradiction fondatrice et la controverse entre Frege et Russell sur le statut de la "variable"*, in [Serfati 2005], 189–193.

When the unknown enters the exponent

In some texts, this 'letter' indicates, for Leibniz, an unknown quantity, usually both unknown and sought after. He states that "it is the very degree of the equation that we ask for," or else "the unknown enters the exponent." Thus, in a letter to Wallis dated May 1697 (GM IV, 23–29)[21], he writes:

> I am the first one, if I am not mistaken, who introduced the exponential equations, in which the unknown enters the exponent. (page 27)

Leibniz pursues, then, his correspondence to Wallis with an occurrence of the "admirable example". However, the same 'letter' can just as well refer – for him – to an indefinite quantity, that is to say, an arbitrary but fixed one (but not necessarily sought for). In this case, Leibniz asserts that "the degree is indeterminate"; it is the *gradus indefinitus* as opposed to the *gradus certus*. In this context, he writes, "the equation has no degree", and in other texts that it has "all the degrees at the same time". One recognizes in this conceptual confusion – so unusual in Leibniz – the intact problem and the contradiction inherent in the aforementioned dialectics of indeterminacy. Let us repeat, this confusion was, at the time, natural and inevitable.

To try to circumvent this difficulty, it sometimes was the case that Leibniz would propose definitions that we might consider "ecumenical." Thus, in a letter to Clüver of 18/28 May 1680, he wrote[22]:

> In the same geometry, it is necessary to know that uncountable problems appear, that I called Transcendent, because they rise above the algebra[23]. And they cannot be returned to algebraic equations, because they are of no degree (that is they are, neither plane nor solid, etc.), either may be of all the degrees at the same time.

Towards letteralized exponentials

Thus, the status of the letter x in a^x will remain naturally ambiguous in Leibniz. This ambiguity falls within the epistemological analysis we have previously exposed: does it symbolize an indeterminate or an unknown? These two possibilities are equally not Cartesian, but they naturally led Leibniz to different conclusions. Let us repeat: Leibniz can consider, initially, that these exponentials characterize equations with indeterminate degree (the Leibnizian *gradus indefinitus* as opposed to the *Cartesian gradus certus*), and are thus likely to exceed all finite degrees (this will be redesigned by Jean Bernoulli); this first solution leads to conceiving of a second origin – more precise symbolically – of the term 'transcendent'. However, Leibniz can also and alternatively consider that he is looking for the value of x. In this case, the "unknown enters the exponent," as in the admirable example.

---

21 See [Breger 1986], 125, n. 38.
22 GP VII, 18–19. This point is raised by {Breger 1986], 125, n. 37.
23 Thus, we discover here, in Leibniz, a still somewhat different meaning of the word "transcendent", always in the exceeding register: problems are *transcendent* when they "exceed" those of algebra.

In these circumstances, I was confronted with uncertainties and fluctuations in Leibniz which were, after all, natural; I therefore preferred to use the term *letteralized* exponentials, which falls within the formal register and does not prejudge any meaning.

On the other hand, we also find in Leibniz exponentials such as $x^{\sqrt{2}}$. If the *gradus* is obviously, here, *certus*, it is however not Cartesian (although it is Newtonian – see above Chap. III). Of these exponentials, Leibniz will form a fully-fledged category, namely the *interscendent* exponentials (see below IV-C)–3).

IV-C3 TOWARDS A HIERARCHY IN TRANSCENDENCE: THE INTERSCENDENT EXPONENTIALS

When the degree of the exponential "falls between" two integers

In his attempt at inventorying the various exponentials, Leibniz brought to light what he considered to be a particular type of exponential expression (and thus of equation); he called them *interscendent* (such as $x^{\sqrt{2}}$) because, he said, their degree "falls between" (in Latin, *inter-cado*) two degrees which are integers (1 and 2 in our example).

One can legitimately consider that this idea had been suggested by the *Epistola Posterior* of Newton (see above, Chap. III-B). One can just as legitimately consider that, for Leibniz, these expressions (and their associated equations) "fall between" the Cartesian algebraic expressions and his own exponential expressions, which were for him *fully* transcendent.

A hierarchy in transcendence?

For Leibniz, the interscendent exponentials might have presented a moment a specific interest because, in this case, he somehow managed to rank the degree of complexity of the exponentials. The expression $x^{\sqrt{2}}$, because it has indeed lost what was, for Leibniz, a constitutive quality of exponential transcendence – namely, indeterminacy – is certainly not an ordinary quantity, but nor is it transcendent.

Thus, Leibniz writes in the *Continuatio Analyseos Quadraturae Rationalium* (GM V, 361–366 = *Continuation of the analysis of the rational quadratures*):

> For example, although $\sqrt[e]{2}$ is not an ordinary quantity, in the case where e is the irrational number $\sqrt{2}$, I shall not however say it is transcendent, but interscendent, because its degree falls between two ordinary degrees (page 361).

Let us recall that, for us today, these exponentials added nothing significant to the mathematical corpus. It is therefore not fruitful to give them a specific status. Leibniz himself ultimately paid little attention to them, and Breger notes (p. 131) that interscendent expressions hardly played a role for him.

Nor did they interest the successors of Leibniz. In the article "Curves", the Encyclopaedia of D'Alembert (page 451) simply mentions, in passing:

[...] the interscendent curves in the equation of which the exponents are radicals, such as $x = y^{\sqrt{2}}$. These two kinds of curves [the letteralized exponentials and the interscendent ones] are properly neither geometrical, nor mechanical, because their equation is finite without being algebraic.

On this point, one can find one of the – very rare – somewhat extended comments in an Emile Turrière's article[24], quoted by Breger. The subject will then fall into oblivion. However, one should note the interest of the voluntarist approach of Leibniz, namely inventorying a new unknown territory (that of the exponentials) by *organizing a hierarchy in it*.

IV-C4 THE EXPONENTIAL UTOPIA

The modern reader might be somewhat surprised when he approaches this 'exponential' part of the work of Leibniz, and he cannot fail to ask some questions: why was Leibniz so very committed (at least initially) to these letteralized exponentials? What is their conceptual importance (if any)? For what specific uses would they be favoured among the non-algebraic functions? For what types of particular uses did Leibniz intend them?

In the 1680s, Leibniz, following the course of his mathematical researches for a time, raised the question of whether one can reduce some quadratures (and, if possible, any quadrature) to letteralized exponentials. This questioning was, at the time, fundamental. It is indeed clear that, if Leibniz had been able to answer positively his own question, he would have completely overtaken Descartes. Indeed, when we remember (see above, Chap. III) the symbolic importance of the Cartesian exponential and of conceptual revolution that it had roused, we understand the desire of Leibniz to provide (with his own exponential) a definitive answer to the issues, which Descartes had not resolved (nor even considered), such as the quadratures and the rectifications.

Leibniz had actually understood (see above, Chap. II-A) that the quadratures of the simplest algebraic curves do not lead to algebraic quadratrices, that is to say, to results expressed with the Cartesian exponentials. However, if one was able to express them with the new letteralized exponentials (i.e., Leibnizian exponentials), everything, then, would have again become harmonious! Thus, it is the possibility of resolving all the analytical quadratures of algebraic curves *via* the letteralized exponentials which was the second decisive argument of Leibniz's approach. On this point, Breger[25] quotes an essay of March – April, 1679, the first lines of which are particularly clear:

We have to see if it is possible to define transcendent quadratrices of analytical curves by means of non-rational exponentials[26].

---

24 See [Turriere 1919]. However, the conclusions of this article are sometimes very questionable.
25 [Breger 1986], n. 47, 126.
26 *Videndum an liceat definire quadratrices analyticarum transcendentes ope Exponentium non-rationalium* (LH 35, XIII, 3, Bl. (f.) 111–112). Dated March 30th (April 9th) 1679.

Such a use of the letteralized exponentials for quadratures – so important to Leibniz – will obviously remain wishful thinking. It is now well understood that letteralized exponentials, whether simple, such as a^x, or more composed, have no relevance in carrying out the integration of algebraic functions. However, in spite of his repeated failures, Leibniz would never completely give up this symbolic approach, which I have here called the *exponential utopia*.

In the continuation of this volume, however, we will see that if his exponentials did not allow for Leibniz the effectuation of analytical quadratures, they were nonetheless of great assistance to him in two other areas: firstly, in the differential calculus (see below, IV-C5)), and secondly in the construction of some transcendent curves (see Chap. V-D). In this way, this amiable utopia was at the same time (somewhat) mathematically reasonable.

IV-C5 LETTERALIZED EXPONENTIALS ARE *SOLUBLE* IN THE DIFFERENTIAL CALCULUS

Let us first return to the original problem of the letteralized exponentials. We have seen how Leibniz allowed himself to "transcend" both Descartes and Newton at the symbolic level by producing an exponential a^x, where x is, in modern terms, the sign of a real number.

The precise definition of $y = a^x$, however, is – to my knowledge – very rarely analysed in Leibniz: such a number is, for him, simply defined by $\log y = x \log a$. To my knowledge, and again, the symbolism $e^{x \log a}$ does not appear in Leibniz, no more than the associated – today usual – power series expansion:

$$a^x = \sum_{n \in \mathbb{N}} \frac{(x \log a)^n}{n!}$$

We have already emphasized that Leibniz obtained very few interesting results regarding the exponential equations. It was also noted that the simplicity of the 'admirable example' could only underline the complete lack of substantial results. In these conditions, what would nevertheless cause Leibniz to try to strengthen his position – should he want to continue this exponential path?

In a manner both spontaneous and skilful, Leibniz then got his calculus to operate on the letteralized exponentials. In other words, in order to handle them mathematically, he did not use the (here inoperative) techniques of algebraic (Cartesian) equations, but rather – and systematically – his own calculus. Thus, only very rarely in Leibniz are there genuine algebraic calculations concerning the exponentials (for example, the standard rules on the exponents, such as $(a^m)^n = a^{m \cdot n}$); however, there are almost always some differential calculations regarding them. Differentiating an exponential is actually a very simple operation: if we do indeed have $y = a^x$, then:

$$\log y = x \log a$$

Therefore, $\dfrac{dy}{y} = dx \cdot (\log a)$ et $dy = (\log a) y \, dx$

Leibniz considered that in fact the exponential is given not by its explicit form (a^x) but rather implicitly, by its differential equation ($dy = (\log a)\, y\, dx$). The latter is indeed linear, on the one hand, and on the other hand – and most importantly – deprived of exponentials and of logarithms as regards the variables x and y. This perspective of an implicit definition, which is easily accessible to the modern reader, was new at the time and would be a continual theme for Leibniz (see the discussion below on this point, in Chapter VI). Such computations, under this exact form, are usual for him; in order to handle the exponential, Leibniz basically uses the differential of the logarithm and not the processed exponential form ($e^{x \log a}$) which is today common.

Leibniz thus considered that, by the strength of his differential calculus alone, he had quite simply removed transcendence. The simplicity of such linear differential computations reassured him and strengthened his conviction that the letteralized exponentials were fundamental elements of his new calculus[27].

An exchange of correspondence with Huygens (1690)

A long exchange with Huygens at the end of 1690 illustrates this conclusion[28]. We will analyse below in Chapter XII (*The transcendence bone of contention between Huygens and Leibniz (1690)*) the detail of this correspondence[29].

In a preliminary letter (of 3(13) October, 1690), Leibniz had solved, by introducing exponentials, a question which Huygens had submitted to him. He found (p. 50):

$$\frac{x^2 y}{h} = b^{2xy}$$

where, he says, the logarithm of b is 1 (for us $b = e$). Leibniz argues:

> I spoke sometimes in the Acts of Leipzig of these equations in unknown exponents; when I can get them, I prefer them to those that are formed only by means of sums or differences. So, they can always be reduced to differential equations, but not vice versa (GM II, 50).

The exponentials are thus, for him, soluble in the differential calculus and the differential equations, but not in the ordinary calculation. This point he would defend for a long time, and with conviction.

---

27 The book of Goldenbaum and Jesseph ([Goldenbaum 2008]), offers a wide variety of interesting studies, both historical and philosophical, of the mathematical thought of Leibniz about the 'infinitesimal differences' and the genesis of his differential calculus.

28 GM II, p. 56. See also Leibniz's response in GM II, pp. 61–62.

29 It therefore belongs to the second phase of the exchanges between Leibniz (who writes from Hanover) and Huygens (who replied from The Hague).

The response to Nieuwentijt (1695)

Leibniz's response to the objections of Nieuwentijt, *Responsio ad nonullas* (...), published in the *Acta* of 1695[30], adds to the subject substantial additional arguments.

Indeed, Leibniz details for his opponent the resolution of a more complicated case: how to differentiate $y = x^v$, where v is a function of x and y?

One has, says Leibniz, $y = x^v$; hence, $v \log x = \log y$.

$$\text{Thus, } v\frac{dx}{x} + dv \log x = \frac{dy}{y}$$

$$dy = d\left(x^v\right) = x^v\left(\frac{v}{x}dx + dv \log x\right)$$

One can thus see, once again, that Leibniz never differentiates the exponential as such, but always its logarithm. Engelsman pertinently notes:

> Exponential curves are those that can be expressed by an equation which involves not only the operations +, -, ×, :, √, but also variable exponents as a^x, x^y etc. Leibniz and John Bernoulli both show how the calculus can be applied to these exponential curves[31].

---

30   GM V, 324. The full title is: *Responsio ad nonullas difficultates a Dn. Bernardo Niewentijt circa methodum differentialem seu infinitesimalem notas; that is, Responses to some objections raised by Mr. Bernard Niewentijt about the differential or infinitesimal method.*
31   [Engelsman 1984], 19–20.

CHAPTER V
GEOMETRY AND TRANSCENDENCE: THE CURVES

V-A THE UNIVERSES OF THE GEOMETRICAL DISCOURSES
OF DESCARTES AND LEIBNIZ

In the inventory of 1679 (cf. aforesaid letter of Leibniz to Tschirnhaus), the question of the transcendent curves had temporarily been set aside, perhaps because of the difficulties which Leibniz had encountered in characterizing too vast a set, with still ill-defined outlines.

Let us recall first that the typology classically held by the Greeks as regards the plane geometry loci was that described by Pappus in his *Mathematical Collection*; it distinguished between three types of loci: *plane* loci (constructible with a ruler and a compass), *solid* loci (constructible using conic sections: ellipse, parabola, hyperbola), and others, which Pappus calls *linear* (*Grammica*).

Descartes and the 'geometrical' curves

Let us then recall that Descartes, who makes a reference in his *Géométrie* to the classification of Pappus (AT VI, 377–378), decided to upset this by reconstructing a typology governed by a new conceptualization *via* the production of a couple of specifically Cartesian and opposed concepts: geometrical curves opposed to mechanical curves[1]. This decision of Descartes was the founding moment of modern geometry. For him, the geometrical curves were only amenable to his (algebraic) calculation; they were defined, on the one hand, by the fact that they possessed an equation, while on the other hand this equation *necessarily* involved a polynomial with two variables $F(x, y) = 0$.

Descartes thus excluded from his geometry all other curves, which he called "mechanical". Descartes wraps this term in a slightly pejorative connotation which he had received from Plutarch (see [Plutarque 1877], 22). He applied it, first of all, to Hippias' quadratrix and the Archimedean spiral – one can find a modern presentation of their respective constructions in [Serfati 1993], 203–204. These two examples represent transcendent curves in the modern sense[2], and Descartes would resume using them unceasingly as models of curves to be rejected in the geometry (see, for instance, A.T VI, 390). Of course, Descartes considered his classification to be assured and definitive. Note that it at least allowed, for the first time in history,

---

1   On the origin of the motivations of Descartes and his approach, see our two papers [Serfati 1993] and [Serfati 2008b].

2   The polar equation of Hippias' quadratrix is $\rho = \dfrac{2a\,\theta}{\pi\,\sin\,\theta}$.

the emergence of the concept (even if it was partial) of 'all the curves' – algebraical, of course … Indeed, up to Descartes, the curves were not seen as forming a class but rather as a group of individuals encountered individually, 'on the ground'[3]. If the classification of Descartes thus definitively settled the issue of a large class of 'geometrical' curves, it nonetheless left unresolved the question of the extension of the vast domain of the 'mechanical' curves.

Leibniz and the 'algebraical' curves

For Leibniz, the challenge was thus to introduce in the new geometrical universe of discourse – non-Cartesian – another still new typology, as he had constructed it previously in the calculation and the symbolism for the expressions and the equations. In a spontaneous way, he began by proposing a new and specific terminology, embodied in the pair of opposed algebraic curves/transcendent curves.

He explained it in a letter to Tschirnhaus (dated 1679 – GBM, 409): "I call Algebraical curves those that Descartes called Geometrical." On the central point of terminology, Breger, who points out that the term 'algebraical' had not appeared in the draft of the letter to Oldenburg (dated 1675), then notes:

> Leibniz introduces the expression 'algebraical' for the curves which are called geometrical by Descartes, so that the word 'geometrical' which confers the legitimacy is used in a broad sense[4].

This comment is relevant: the change of terminology will indeed allow Leibniz to accept that some of the 'non-algebraical' curves (in his view, that is to say, in the sense that Descartes gave to 'non-geometrical') are nevertheless, for him (Leibniz), 'geometrical' and thus, in some sense, acceptable to him. However, as Engelsman rightly points out, the difficulty for Leibniz arises from the manner in which the concept of the curve was – at the time – badly defined. He writes: "By consequence it was not at all clear how many non-algebraic curves did in fact exist"[5]. This point will be central to the strategy of a rational inventory of Leibniz.

The remark by Engelsman is relevant regarding the concept of a curve in the XVII century. This concept appears, moreover, and even nowadays, as multifaceted, and it is not easy to elaborate. We present below, in *Chapter XV*, a sketch of a correct definition. Engelsman concludes:

> Thus the universe of discourse[6] at the end of the XVIIth century consisted of three classes of curves, the algebraical, the exponential, and those transcendent curves that could be expressed by means of integrals (Engelsman, *Introduction*, page 20).

However, this final comment concerning the conceptions of Leibniz seems somewhat premature. Actually, the conclusion of Leibniz concerning the issue of a sys-

---

3   This is pointed out by [Engelsman 1984], *Introduction*, 19.
4   [Breger 1986], 123.
5   [Engelsman 1984],18.
6   This relates to the geometrical discourse. For that of the calculation, it is obviously different.

tematic inventory of the vast universe of non-Cartesian curves was somewhat more pragmatic. He first discovered, within the non-algebraical curves, extraordinarily diverse backgrounds. As a first step, Leibniz had recognized the superiority of the symbolism (that is to say, the importance of the equations, whether Cartesian or exponential) to represent curves. However, he also realized that the objects which we can enclose into an equation, and those which we can build do not have the same extension. Breger pertinently notes:

> Having gone beyond the Cartesian boundaries, Leibniz is confronted with this problem: what can be expressed in an equation and what can be built are no longer identical. In the following years, it becomes increasingly clear that Leibniz gives primacy to the equation as a criterion of decision; his interest remains relatively restricted for new instruments of construction by means of which might be defined what is geometrically authorized ([Breger 1986], 123, & n. 21–23).

In the sections which follow, I will thus distinguish the diverse concepts of curves recognized by Leibniz. First, what in modern terms we call 'parametrically defined' curves (such as the cycloid or the spiral), then the envelopes (and the evolutes), and finally the 'physical' curves. The case (more generally) of curves defined by an integral is discussed later, in *Chapter VI: Quadratures, Inverse Method of Tangents and Transcendence.*

V-B LEIBNIZ'S CRITIQUE OF DESCARTES' GEOMETRY 'OF THE STRAIGHT LINE' AND THE PARAMETERIZED CURVES.

For Leibniz, a curve that does not have any equation, whether Cartesian (i. e., via a polynomial) or 'Leibnizian' (i. e., defined by an integral or via letteralized exponentials), must nevertheless always be considered, both for its interest in itself (it was often very well known since antiquity) and for when it appears as the quadratrix curve of a given curve. Given these conditions, such a curve appears as a purely geometrical object and can be obviously obtained only by a construction that is also geometrical.

Leibniz thus considers that the cycloid and the trochoid, etc., rejected by Descartes outside his geometry within the 'mechanical' domain, are curves equally acceptable as the circle. They are not, however, amenable to an equation in the style of Descartes. Very early on (from 1674), Leibniz thus denounces another insufficiency – a major one according to him – of Cartesian analysis, which does not allow the inclusion in an equation of so standard a curve as the cycloid[7]. In a letter to Mariotte dated October 1674, very clearly on the subject, Leibniz details the base of his critique of the Cartesian notion:

> Mister Descartes worked, after Vieta, upon reducing the equations of Geometry to the resolutions of the Equations, the calculation of which is completely Arithmetical. But he, just like Vieta, was only interested in Rectilinear Questions, that is to say, those in which one only seeks (and also, one supposes) the magnitude of a few straight lines or rectilinear figures; to which are indeed reduced all plane problems, as well as the solid, the sursolid, etc. But we must admit

---

7　On Leibniz and the transcendent curves, see also [Grosholz 2007], 204–224.

that the Curvilinear Problems are of an entirely different nature, and that we can say that they are neither plane, nor solid nor sursolid, but of an infinite degree, if we want to solve them by equations (A III, 1. N.38).

Descartes' practice of the construction of curves, 'linear, pointwise'

This early correspondence, which criticizes the 'linear' perspective of Descartes's geometrical thought, may raise certain questions among contemporary readers. What, exactly, are these straight lines for Descartes? Leibniz's statements, however, are well-founded. To estimate them, we first highlight Descartes' conception of the elaboration of the equations (obviously algebraic) of his curves (see [Serfati 2008b], 31–34). There follows an example of Descartes in a well-known extract of *Géométrie* (Book II):

> (...) I think the best way to group together all such curves and then classify them in order is by recognizing the fact that all the points of those curves which we may call 'geometrical', that is those which admit of precise and exact measurement, must bear a definite relation to all points of a straight line, and that this relation must be expressed by means of a single equation (AT VI, 392).

This passage is somewhat convoluted, although it is epistemologically decisive. It requires a precise analysis[8]. What Descartes is claiming with the expression "bear a definite relation to all points of a straight line," has indeed been poorly understood by some commentators. Descartes actually wants – each time – *to elaborate his equations of curves* by considering, *for every fixed y,* the solutions x of an equation of the type:

$$a_0(y) x^n + a_1(y) x^{n-1} + \ldots + a_{n-1}(y) x + a_n(y) = 0 \ (= F(x, y)) \ (1)$$

where a_j denotes real polynomials (F is thus a real polynomial with two variables). This procedure is particularly clear in the case of the calculation of the equations of the Pappus curves[9].

Thus, what we described as 'for every fixed y' ultimately means that a certain point (with an abscissa y) describes a straight line. The "relation to all points of a straight line" – a genuine leitmotif of the *Géométrie* – must be understood precisely as the recognition of the dependence (below indicated via the function H_y) embodied in what is now identified as one of the two partial applications associated with a polynomial with two variables and its roots, namely $x \rightarrow H_y(x) = F(x, y)$. It is indeed sufficient to return to the relation (1) and to posit:

$$F(x, y) = H_y(x) = 0 \ (2)$$

For Descartes, the central question was therefore to *construct*, for each fixed y, the roots x of the polynomial H_y – and it is exactly this 'linear' conception that the critique of Leibniz is concerned with.

---

8   See our developed study in [Serfati 2008b].
9   *Ibidem*, 27–34.

As can be seen in (1), Descartes authorized only the equations of plane curves defined, either implicitly ($F(x, y) = 0$) or explicitly ($y = G(x)$, which is the same) – this is the essence of his method – but not what we now call the 'parameterized' curves. Descartes thus excluded completely some very common curves (as Newton would underline so insistently).

Leibniz and the parameterized curves

We take the example of the cycloid, which at Leibniz's time was only amenable to a geometrical construction (though it is very simple construction, that is to say, the locus of a point attached to a circle which rolls without slipping on a given straight line). Today we usually define the cycloid in the calculation by the following parametric representation:

$$x = a\,(t - \sin t)$$

$$y = a\,(1 - \cos t)\ (t \in \mathbb{R})$$

Such a parameterization is now regarded as the 'natural' form of representation of the curve. However, it is highly difficult to represent it in Cartesian terms such that $y = G(x)$ or else $F(x, y) = 0$, with G (or F) as an algebraical function[10].

Leibniz thus understood, confusedly, that something essential was missing in the Cartesian 'linear' method without being able to completely imagine the modern, subsequent solution, which would result from the development of the theory of parameterized curves. The same conceptual obstacles apply to the Archimedean spiral and Hippias' quadratrix – both are also equipped with very simple geometrical constructions[11]. The fact that Leibniz wanted to use parametric coordinates (or 'curvilinear' coordinates) remains controversial among historians and commentators. This point is emphasized by Breger (p. 124, n. 31). From the letter to Mariotte above, one can nonetheless understand to what a large extent the idea was present in him. On the other hand, the last pages (GM VII, 292–299) of *Specimen Geometriae Luciferae* (1695) present an impeccable construction from Leibniz of plane curves defined by the composition of two movements – horizontal and vertical – and thus a qualitative conceptualization of parameterized curves and their main properties (increasing or decreasing aspects, inflection points, etc.).

---

10   Nevertheless it is possible by restricting to a single given arch of the cycloid – however with some difficulty: It requires introducing a new transcendent function in the modern sense, that is to say: $x \to \mathrm{Arc\ cos}\ x$.

Indeed, with $a = 1$, one has $y = 1 - \cos t$, and $t = \mathrm{Arc\ cos}\ (1\text{-}y)$, if $t \in [0, \pi]$.
Therefore $\sin^2 t = 1 - (1-y)^2 = 2y - y^2$, and:

$$x = Arc\ \cos\ (1 - y) - \sqrt{2y - y^2}$$

Neither Leibniz, nor Newton could use such a practice of calculation.

11   See our presentation in [Serfati 1993], 203–204.

V-C EVOLUTES, ENVELOPES
AND THE GENERATION OF TRANSCENDENT CURVES.

The evolute of an algebraical curve is – usually – a transcendent curve

The instance of what is now called the *envelope* of a one-parameter family of plane curves is an example of a completely different nature, of a curve which, though very usual in geometry, is not necessarily provided with an explicit equation – although, obviously, it could be equipped with it (see below, *Section V-F: Mathematical Complements*). In addition, if it admits such an equation, it is obtained not using an 'ordinary' geometrical construction as in the previous section, but by the use of the differential calculus (see *ibidem* V-F). Obviously, this type of curve was thus exemplary in a double title in the eyes of Leibniz.

The theory for determining the envelope of a one-parameter family of straight lines naturally preceded – historically – the more general issue of the envelope of a family of arbitrary plane curves (still with one parameter)[12]. As to the straight lines, Leibniz does not use the modern terminology (i.e., 'one-parameter family'). For lack of a proper symbolic setting – difficult to conceive in his time – he uses a geometric representation, explaining that the family under consideration is composed of *curves*, of which he says they are *given by position* or else, *ordered: lineas ordinatim ductas,* or *lineas ordinatim (positione) ductas* (GM V, 267). He considers, in fact, that the parameter behaves as a point describing a certain fixed straight line or curve – defined as the 'ordinatrix'. To every position of this point on the ordinatrix there corresponds a curve of the family (see GM V, 267) [13].

Regarding the envelopes of straight lines, the theory had been widely introduced by Huygens in *Horologium Oscillatorium* (Paris 1673); in this text, he establishes the theory on the paradigmatic case of *the evolute* (D) of a given curve (C), that is the envelope of a particular one-parameter family of straight lines, constituted by the normals to (C) (this curve is generally supposed to be algebraic); the parameter in question, here, is that which fixes the position on (C) of the current point. The correspondence which, to a curve (C), associates its evolute (D) is therefore 'from curve to curve'. However, one can easily determine that the evolute of an algebraical curve is not algebraical but actually transcendent – even for very simple cases. As such, the evolute of an algebraic curve as simple as an ellipse is transcendent, namely, the image of an astroid by an orthogonal affine transformation[14].

---

12   To illustrate this, Leibniz often takes the classic example of what is now called *caustics by reflection* in relation to some given mirror. In this case, the reflected rays obviously constitute the one-parameter family of straight lines (GM V, 267). The position of the contact point on the mirror defines the parameter. The mirror is thus the ordinatrix.

13   See [Engelsman 1984], 23–24.

14   For an ellipse (E) given in canonical equations by $\frac{x^2}{a^2} + \frac{y^2}{b^2} = 1$ $(a>b>0)$, the evolute (D) has

for parameterization: $x = \frac{c^2}{a}\cos^3 t;\ y = \frac{c^2}{b}\sin^3 t\ (t \in \mathbb{R})$, where c is the focal distance:

$c^2 = a^2 - b^2$. This transcendent curve is the image by affinity of the astroid $x = K\cos^3 t,\ y = K$

$\sin^3 t$, where $K = (c^2/a)$. An implicit equation of (D) is $\left(\frac{ax}{c^2}\right)^{\frac{2}{3}} + \left(\frac{by}{c^2}\right)^{\frac{2}{3}} = 1$. If the ellipse is a

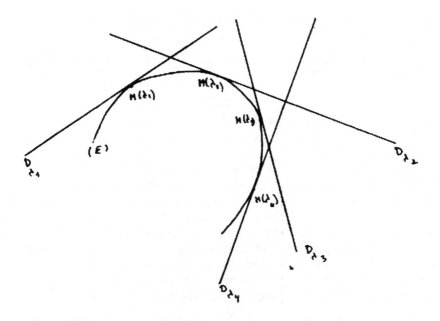

Quadratures and evolutes are processes
that are at the same time reciprocal and similar

Epistemologically speaking, Leibniz discovered, here, a remarkable mathematical situation, as it is at the same time both the analogue and the reciprocal of that of the quadratures: starting from algebraic curves, the new operation, namely the passage from a curve to its evolute, allows the construction of transcendent curves in a relatively systematic way. This is what is also forming the quadrature process – the significant difference being that the quadrature falls within the register of summation, and the evolute within that of differentiation, namely the two basic reciprocal operations invented by Leibniz. Moreover, Leibniz never failed to observe, since the beginning of his mathematical works, that this reciprocity between theoretical concepts (differentiation/summation) was covering, in truth, a profound asymmetry in practice: it is, in general, much easier to differentiate than to sum (i.e., to integrate). In the present case, it is easier to search for the evolute of a curve than one of its involutes (that is to say, a curve of which it is the evolute – there are infinitely many such curves. See *Complements*, below). It is therefore not surprising that Leibniz was able to consider the determination of the evolutes as a (new) paradigm for the systematic construction of transcendent curves starting from algebraical

circle ($a=b$), the evolute is obviously reduced to a point (the centre). The evolute of the hyperbola is transcendent. The evolute of the parabola is an algebraic curve.

curves. This conceptual mechanism (which remains, to my knowledge, very rarely mentioned by commentators) is in contrast clearly highlighted by Breger:

> Occasionally, Leibniz even wondered if one could not build all the transcendent curves as some evolutes ([Breger 1986], 124 and n. 29).

'On the art of the discovery in general':
Leibniz and the need for reasoned inventories

In support of the above remark, Breger cites a surprisingly early text (1678) of Leibniz, in the edition of Couturat, namely the *Arte Inveniendi in Genere* (On the art of the discovery in general)[15]. This text – the manuscript of which is among those that Leibniz covered the most with notes in margin[16] – is of great importance epistemologically. Leibniz indeed exposes the theoretical conditions of a "perfect invention", beginning with the possibility of producing an *a priori* conceptualization of a systematic inventory of the unknown object. One can well-understand to what extent this epistemological issue was maintained in Leibniz by his desire for an inventory of the non-Cartesian world, namely the transcendence:

> A method of invention would be perfect if we could plan, and even demonstrate, even before entering the subject, which are the ways by which we shall reach its completion (p. 80).

The text is full of relevant observations on this practice, which we cannot reproduce in their entirety here. Furthermore, and for example:

> A method of inventory is not perfect if you do not determine some report, for example the list of transcendent curves defined as the evolutes of either one or several curves, i.e., the search for the fact that any transcendent curve can be described by means of one or several evolutes. Indeed, if a transcendent curve cannot be obtained as the evolute of a single algebraic curve, it would be necessary to look for if it cannot be by means of two algebraic curves, or by three, etc. Thus, we get a genuine Method. (AVI 4, N. 29, 82–83) (= [Couturat 1903], 164).

By this important text, Leibniz clearly makes us understand the esteem in which he holds the evolutes and to what extent he hopes that the evolutes of the algebraic curves are a paradigm for obtaining transcendent curves, i.e., that any transcendent curve is a certain evolute. He simultaneously sketches a reasoned inventory of transcendent curves: such a curve being given, it can be generated as the evolute of an algebraic curve D1. If it is not the case, it can be as the evolute D2 of the evolute D1, etc. Thus, a constant, central epistemological approach of Leibniz was the promotion, in a new science, of reasoned and exhaustive inventories[17].

---

15 The text is now also available in the edition of the Academy, AVI, 4, N. 29, 79–83, as well as in [Couturat 1903], 161–166. See the commentary and English translation in [Dascal 2008b], chapter *Towards an invention for discovery*, 94–98.
16 This is pointed out in [Dascal 2008b], 94.
17 This is pointed out by [Dascal 2008b], 103, who pertinently makes a reference to Breger.

Leibniz and the envelope of a family of curves

We will continue this study with another important text, and much later, of Leibniz. It is a well-known article, devoted to the general definition of the envelope of a family of curves (and not only of straight lines), published in the *Acta* (1692) (GM V, 267), and whose title is a true programme:

> *De linea ex lineis numero infinitis ordinatim ductis inter se concurrentibus formata easque omnes tangente, ac de novo in ea re analyseos infinitorum usu*, i. e., "On the curve constructed from an infinite number of ordered and intersecting curves, which is tangent to all, and on a new application in mathematical analysis that follows."

This text is equally remarkable in terms of its historic aspects as it is by its theoretical conceptions; it marks the creation by Leibniz of the *general concept of an envelope*. He begins by proposing a detailed history of the issue of the envelopes of a one-parameter family of straight lines, quoting Huygens and the evolutes, and Tschirnhaus and the caustics; these are indeed two geometrically important examples, whose relation to physics is evident. In accordance with its permanent infinitesimal perspectives, Leibniz first considers what is called, in modern terms, the 'characteristic point of a normal' – that is to say, the point at which it touches the envelope – as the meeting point of two infinitely close normals and which he supposes to be unique. According to an important theorem of differential geometry – which Leibniz knows well – it follows, then, that this characteristic point is also the centre of the osculating circle to the initial curve at the point concerned (see below, *Mathematical Complements*).

However Leibniz also understood that all that had been said about one-parameter families of *straight lines* might as well be written for one-parameter families of *curves*. Thus, he concludes:

> But while this only applies to straight lines, you must know that something similar happens in the case of curves (GM V, 267).

Leibniz next reveals, according to the title of the article, the general theory of the envelope of a one-parameter family of *arbitrary* curves (see below, V-F). He then considers, by extension, that the characteristic point of a curve of the family – that is the point (always supposed to be unique) at which it touches the envelope – is the meeting point of two infinitely close curves.

Variables or constants? Exchanging interpretations

Note also that this theory of envelopes, which is still taught as such to students, presents an epistemological aspect that is both as important as it is unexpected at the symbolic level – well-identified by Leibniz himself – namely, the exchange of interpretations about the letters: from constants, they become variables, and vice versa. Indeed, this is true within the same symbolism, namely that of the equation of the family of curves under consideration.

Leibniz thus explains, and at the same time, the principle of the keys of interpretation and – he says – the absolute necessity, in this case, to change them. On the one hand, he says, there are in the ordinary of the differential calculus, parameters (*parametri*, 268), or constants[18], which we are used to designating by *a* and *b*. These are then fixed (*indifferentiable*). On the other hand, there are coordinates, such as the abscissa and the ordinate, which are related to each other (*gemina*) and are thus variable (Leibniz says *differentiable)*, and which we are accustomed to indicate by *x* and *y*. In the case of envelopes, however, the exact opposite will occur: the parameters must be considered to be variable, and therefore they can be differentiated, while the coordinates are fixed. This is exactly equivalent to saying that we have inverted the keys of interpretation. Thus, Leibniz rightly claims the full procedure:

> But in our present calculation, we do not seek the tangents to a unique curve at an arbitrary point, but a unique tangent to an infinite number of ordered curves given in position; we are therefore seeking the point of contact (…) and *x* or *y* is unique (…); but at least one of the parameters *a* and *b* must certainly be (…) differentiable[19]; so, by making it vary, one passes, among the ordered curves, from one given curve to another (GM V, 268).

Next, the ultimate work for Leibniz was to try (naively!) to display a category of super-constants (*constantissima*; GM V, 269), which are supposed to never again be able to be differentiated[20]. In a subsequent text[21], *Monitum De Characteribus Algebraicis* (1710), Leibniz, certainly inspired by his invention of envelopes, would theorize further what we have called the 'exchange of interpretations'. This systematic change in the status over the same symbolism – quite exceptional in the history of mathematics – is consubstantial with the very method of determining envelopes invented by Leibniz (N.B. we have analysed at greater length this decisive symbolic issue in another work[22]).

We can finally underline that L'Hôpital devoted a whole chapter of his *Analyse des Infiniment Petits* (1696) to the theory of evolutes (Section V, p. 71, *Usage du Calcul des Différences pour Trouver les Développées*). Another chapter is dedicated to caustics by reflexion. Unsurprisingly, for l'Hôpital, both issues are a matter of the differential field alone (and not of the integral calculus).

---

18 More precisely, Leibniz says that they are "constant magnitudes" (*rectae magnitudinae constantes*).

19 Leibniz here uses the adjective *gemina*, which we have not translated and which might mean that the parameters *a* and *b* are connected. This, however, is a rare special case in practice. Rather, one should consider that Leibniz conceived, here, the parameters *a* and *b* as being *interdependent*, by pure analogy with the coordinates *x* and *y* – they are actually connected by the equation of the curve. As such, these are actually the coordinates (*coordinatae*) according to the term invented by Leibniz in this text (GM V, 267).

20 It is thus to Leibniz that we owe the introduction of the terms 'variable' (in this indeterminate connotation), 'constant' and 'parameter' (still in this sense of an arbitrary but fixed magnitude). Moreover, and although in retrospect, we understand today how, in an article devoted to this very specific technique of the determination of envelopes (in modern terms, one 'differentiates with respect to the parameter'), it was really essential for him to elaborate upon the pair of opposite constants / variables, even if at no time would he go beyond Vieta in the precision of his definitions. On this question of the 'genesis of the variable', see our article '*La dialectique de l'indéterminé, de Viète à Frege et Russell*', in [Serfati 1999].

21 GM VII, 218.

22 [Serfati 2005], 172–173.

V-D TRANSCENDENT CURVES AND 'POINTWISE' CONSTRUCTIONS

In the early 1690s, Leibniz was suddenly interested – with great obstinacy and strength, and in a way that may at first sight seem surprising for the time – in those curves which he would label *percurrentes*, that is to say, a category of transcendent curves which can be, according to him, 'constructed' by means of ordinary geometry. The question of the status of these percurrent entities (equations and / or curves) and of the distinction between the percurrent and the transcendent undoubtedly marks a shift that occurred over time in the position of Leibniz. Breger (p. 129, notes 61 to 63) also highlights an ambiguity between Leibniz and John Bernoulli on the definition of the term. In an extensive commentary and abundantly supported by notes, he writes:

> Note that Leibniz, even after his introduction of the transcendence, was still interested in constructing transcendent curves in a traditional way. Leibniz labelled 'percurrent' these curves for which one could construct, by means of a single transcendent quantity, as many points as needed, in a traditional way. The proof that all transcendent curves in this sense are "percurrent" is, for Leibniz, "the first of his wishes"[23] – yes, it may even be "the peak of the geometry of the transcendence"[24]. ([Breger 1986], 129).

An exemplary debate between Leibniz and Bernoulli concerning percurrence is detailed below in *Section V-E Percurrence and transcendence, Leibniz and John Bernoulli, the story of a misunderstanding*).

The equilibrium position of an inextensible heavy thread

The above extract from Breger is illustrated (in footnotes) by very numerous references to the work of Leibniz[25]. We will resume with some of these texts, the detailed analysis of which we will propose. The case began in the early 1690s with the catenary, which was the object – on Leibniz's part – of a very important number of well-documented and voluminous articles, and which we briefly examine below. This is a problem whose history is now very well-known and studied, particularly by Montucla who, in his *Histoire des Mathématiques* ([Montucla 1799, II], 446–450), has provided both a history and an interesting mathematical analysis. In the context of this work, we focus only on the impact of this issue on the views of Leibniz regarding transcendence.

The problem had already been raised by Galileo – who had given an invalid solution – and then taken up again by James Bernoulli in 1690. While the latter did not have the solution, he communicated the question to Leibniz as a challenge, and the latter was the first to arrive at a correct solution. Afterwards, Leibniz organized (still in 1690) a public competition concerning the catenary, to which, as he writes,

---

23 *Estque id ipsum ex meis desideratis unum* (…), GM III, 207.
24 *Id fateor, fastigium foret Geometriae transcendentis*, GM III, 217.
25 (GM III, 201, 223, 319); (GM III, 139, 144)]; GM V, p. 244, p. 255, p. 264–265, p. 311–312.

only Huygens and the Bernoulli brothers answered within the deadline[26]. As he explains it to Huygens in a letter of February, 1691, Leibniz considered, however, that the solution which James Bernoulli had "finally" given would inevitably be inspired by his own (i. e., Leibniz's) and his infinitesimal methods:

> Mr Bernoulli also found, finally, the curve for the chain. I believe that the knowledge of my calculation will have helped him a little because, although this problem is not among the most difficult, I imagine that it is not so easy to succeed in it without having something equivalent to this calculation (GMII, 84).

The problem is stated as follows: given two points A and B of the same vertical plane (not situated on the same vertical line), determine the equilibrium curve of a weighing thread hanging between these two points, A and B, and placed in a uniform gravitational field. The thread is supposed to be flexible, infinitely thin, homogeneous and inextensible. Here, we see a problem of *mixed mathematics*[27], combining mathematics and physics (a hybrid discipline which would be widely continued a little later on, at the end of the century, for example with the problem of the brachistochrone[28]). The problem amounts in the following one, with a somewhat more mathematical statement: determining the curve containing A and B, such that the centre of gravity of the *material* arc $\overarc{AB}$ (homogeneous, and as previously defined) is the lowest possible (i. e., such that its ordinate is the minimum).

Some elements of hyperbolic trigonometry

We will now briefly recall the solution of the problem in modern terms. First, define for any real number x the **hyperbolic cosine** function (in symbols: ch) by:

$$ch\ x = \frac{e^x + e^{-x}}{2}\ x$$

As with all the functions of hyperbolic trigonometry, this will later (in 1770) be identified by Lambert[29], following Ricatti. For Leibniz, it appeared to be directly related to the "logarithmic" (i. e., the 'Leibnizian' exponential). Let us recollect that both functions (the hyperbolic cosine and the hyperbolic sine) receive their names from the fact that they parameterize the equilateral hyperbola of the equation $X^2 - Y^2 = 1$[30], in the same way as the usual (i. e., circular) cosine and sine parameterize the circle of the equation $X^2 + Y^2 = 1$. Recall also that the hyperbolic cosine function is infinitely differentiable (i. e., it is a C^∞-function) and that its first derivative is

$$sh\ x = \frac{e^x - e^{-x}}{2}$$

---

26 Leibniz pressed Tschirnhaus to participate in it, but he refused (GM II, 84).
27 This point is rightly emphasized by [Parmentier 2004],191.
28 See AIII, 6 N. 243 & AIII, 7, N.17 et N. 29.
29 See [Lambert 1770].
30 Indeed, we have for all x, $\mathrm{ch}^2 x - \mathrm{sh}^2 x = 1$; this relation is said to be a *fundamental of the hyperbolic trigonometry*. It is the hyperbolic counterpart of the circular relation $\cos^2 x + \sin^2 x = 1$.

This is the *hyperbolic sine*. We might also note the existence of an inverse function – only partial[31] – for the hyperbolic cosine. It is denoted *Arg ch*, and is defined for $y \geq 1$ by:

$$y = ch(x) \Leftrightarrow x = Arg\ ch(y) = \ln\left(y + \sqrt{y^2 - 1}\right)$$

It is also a C^∞-function for $y > 1$[32]; its first derivative has the value $\dfrac{1}{\sqrt{y^2 - 1}}$. This result is historically important because it is from this derivative alone (more precisely, from $\dfrac{a}{\sqrt{y^2 - a^2}}$, i.e., from the derivative of $y \to a\ Arg\ ch\dfrac{y}{a}$) that Leibniz will recognize the inverse function and, at the same time, the hyperbolic cosine as the solution[33]. From the above, it follows that if one has in modern terms the relation (1) below:

$$x = \int \frac{a\ dy}{\sqrt{y^2 - a^2}}, \text{ then:}$$

$$y = a\ ch\frac{x - x_0}{a} \quad [34]$$

Leibniz will recognize the differential status (that is to say, for us today, the derivative) without ever explaining the function itself. Here, we find a constant of his thought on the definition of curves (see above, IV-B), which he identifies, not in 'Descartes' way by all of their points, but by all their tangents, that is to say via their differential equations.

The catenary

In these conditions, we can now call *catenary* the plane curve of the equation:

$$y = a\ ch\frac{x}{a}$$

In his 1691 article, Leibniz says that the curve solving the problem of the equilibrium of a weighing thread is an **arc of catenary** passing through the points A and B. Regarding this problem of equilibrium, the intuitive ideas of the early XVIIth century were rarely fruitful; as usual, Leibniz proposes a detailed history of these (GM V, 243–244). Thus, in 1638, Galileo, who had been interested in the issue, had believed that the solution was an arc of a parabola[35].

---

31 This actually applies to the restriction to IR^+ of the function ch (i.e., $x \geq 0$). Consequently, Arg ch $t \geq 0$ for every $t \geq 1$.
32 The function is not differentiable for $y = 1$.
33 See [Keller 2009]. Also [Parmentier 2004], 190–191.
34 x_0 is a constant of integration.
35 Roughly speaking, a parabolic arc is 'sharper' than an arc of catenary.

We should note that, in his papers of 1691 (articles and letters), Leibniz, even if he describes with great precision the mode of construction of the catenary (and secondarily its properties) – he presents it as the solution curve of the problem – provides very little information as to how he achieved it. Later, a text by John Bernoulli, *Lectiones Integralis Calculi* (1691–1692) shed light upon this point. His method for elaborating an equation uses, very simply, the laws of statics (see [Keller 2009], 3–4). In addition, Montucla, who was naturally interested in the question, "believed he could not avoid" an explanation on the subject[36]:

> We believe we cannot avoid putting, here, the readers (i.e., geometers) a little on the way towards the solution of this curious and difficult problem. We shall borrow, for this purpose, the subtle analysis which was given by Mr John Bernoulli in the *Lessons of Integral Calculus*. (pp. 446–447).

He then presents, in detail, Bernoulli's method (pp. 447–448). This helps us to understand the way in which Leibniz very probably proceeded. We will give here only a summary of the mathematical part of this reconstruction[37].

The method is fundamentally based on the resolution of the differential equation of the required curve, which is, in modern terms, of the type $\dfrac{dy}{dx} = \dfrac{s}{a}$ (2), where s is the curvilinear abscissa. This leads to $dx = \dfrac{a\, dy}{\sqrt{y^2 - a^2}}$ (it is relation (1) above)[38].

In his commentary, Montucla recognizes that the curve solution is transcendent:

> The catenary is, as we see, a mechanical or transcendent curve since its construction involves the quadrature of the hyperbola (p. 448).

The catenary was a curve provided with various facets, endowed with an undeniable rhetorical value for Leibniz)[39]. On the one hand, its definition is equivalent to that of the logarithmic curve, which is associated with the quadrature of the hyperbola. Yet it also has a physical construction, by means of a thread. Last, but not least, it will be – for him – the constituent model of percurrent curves (see below).

---

36 [Montucla 1799, II], 446–448.
37 It is analysed in detail, in modern terms, by Keller in *Le calcul différentiel de Leibniz appliqué à la chaînette* = [Keller 2009], 3–6. See also [Parmentier 2004], 189–191.
38 $\left(\dfrac{dy}{dx}\right)^2 = \dfrac{s^2}{a^2}$ and $\left(\dfrac{ds}{dy}\right)^2 = \dfrac{(dy)^2 + (dx)^2}{(dy)^2} = 1 + \left(\dfrac{dx}{dy}\right)^2 = 1 + \dfrac{a^2}{s^2}$

By choosing $\dfrac{ds}{dy} = \dfrac{\sqrt{a^2 + s^2}}{s}$, one then gets the differential equation with separate variables:

$dy = \dfrac{s\, ds}{\sqrt{a^2 + s^2}}$ et $y - y_0 = \sqrt{a^2 + s^2}$.

By choosing 0 for the value of the constant of integration y_0, one has:

$s = \sqrt{y^2 - a^2}$ et $\dfrac{s}{a} = \dfrac{dy}{dx} = \dfrac{\sqrt{y^2 - a^2}}{a}$

from which follows relation (1) *supra*.
39 About various constructions of the catenary, see [Grosholz 2007], 222–223.

"On the curve formed by a thread ...".
The article of 1691

Among the large number of Leibnizian texts we are concerned with, I will choose, in the first place, the *De Linea in quam flexile se pondere proprio curvat, ejusque usu insigni ad inveniendas quotcunque medias proportionales et logarithmos*, which appeared in the *Acta Eruditorum* of June 1691 ('On the curve formed by a thread as a result of its own weight, and on its outstanding use to find – as many as necessary – proportional means and logarithms') (GM V, 244–247). Leibniz first outlines the exemplary nature of the curve, which he lists without hesitation among the transcendent ones:

> I indeed discovered that this curve, the easiest to be realized, and at the same time the most useful by its applications, is unrivalled among the transcendent ones. Because we can describe it and realize it without any effort, by means of the suspension of a thread (or better, of a *small chain*) which is inextensible, and so by a certain construction of a *physical type* (...) Descartes, who had remarkable ideas about the reform of sciences, after undertaking calculations and experiments, had excluded the parabola, but without proposing in its place the real curve (...) Having myself tried the experience to please him [i. e., Bernoulli], I have not only succeeded, but I also was the first one, if I am not mistaken, to have solved this famous problem, and at the same time I discovered remarkable applications for this curve (GM V, 243–244).

It is thus necessary to go beyond the simple – and false – ideas which one might spontaneously jump to (e.g., the solution should be a parabola, etc.), and conclude that the catenary is actually an invention of Leibniz. The first part of the article focused on the physical construction of the curve – by means of a thread. Leibniz then develops what is for him the real challenge of his research on the catenary, namely "to determine geometrically as many points of it as we wish." He thus writes:

> This is a curve so constructed Geometrically, without the help of any thread or chain, and without using quadratures; it is the type of construction that we can consider in my opinion as the most perfect and most in compliance with the Analysis as regards the transcendent curves (GM V, 244).

In modern terms, the active ingredient of the proof of Leibniz on this fundamental point must be analysed – in summary – as follows: to construct pointwise the curve of the equation $y = \dfrac{e^x + e^{-x}}{2}$, it is enough to know how to do it for the curve of the equation $z = e^x$; this one, which for us is the exponential curve, is constantly designated by Leibniz by the name of its inverse function, that is to say the logarithmic function (often, he simply refers to "the logarithm") (GM II, 96).

The geometric-harmonic construction and the letter to Huygens of July 1691

Leibniz resumes his argumentation in a letter to Huygens, very slightly post-dating the publication of the article (it is dated 14 (24) July, 1691 (GM II, 95–98)). He comments on his text published in the *Acta* of June. Both of the arguments of Leibniz, firstly in the article *De Flexile*, and secondly in the letter to Huygens, are similar.

To construct 'the logarithm' by means of ordinary geometry, Leibniz uses a coupling that he favours between arithmetic means, in abscissas, and geometric means (which he calls "proportional means"), in ordinates. We will describe the process in modern terms via an internal composition law, which we denote '*'. Given two points P and Q of the plane, then with the coordinates (x_P, y_P) and (x_Q, y_Q), Leibniz defines the point $T = P * Q$ by its coordinates:

$$x_T = \frac{x_P + x_Q}{2} \qquad y_T = \sqrt{y_P \cdot y_Q}$$

The point T is clearly constructible with a ruler and a compass from P and Q, that is to say by "ordinary geometric means." Leibniz then considers four points O, A, Z and W of the plane (arranged as in Figure 23, below = GM II, 96) such as $OW = OA$, and fixes the ratio $\lambda = \dfrac{OA}{WZ}$.

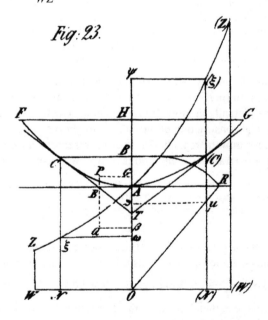

Fig: 23.

We will explain the method in modern terms, taking W as the origin of an orthonormal system (Leibniz does not use any coordinate system). The points O, A and Z have the coordinates:

$$0 \, (\alpha, 0), A \, (\alpha, \alpha), \text{ and } Z \, (0, \beta), \text{ with } \frac{\alpha}{\beta} = \lambda .$$

Leibniz constructs, then, the point $Z * A$, denoted $_1\zeta$ in the article (and not in the letter), which thus has the coordinates $\left(\dfrac{\alpha}{2}, \sqrt{\alpha\beta}\right) = \left(\beta\dfrac{\lambda}{2}, \beta\sqrt{\lambda}\right)$. Next, by dichotomy, he does the same for the couples $(Z, {}_1\zeta)$ and $({}_1\zeta, A)$, by constructing – always by ordinary geometry – two new points $Z^*(_1\zeta)$ and $(_1\zeta)^*A$, which have for their respective coordinates $\left(\beta\dfrac{\lambda}{4}, \beta\lambda^{\frac{1}{4}}\right)$ and $\left(\beta\dfrac{3\lambda}{4}, \beta\lambda^{\frac{3}{4}}\right)$. Leibniz, naturally, then iterates the process. Between two consecutive points already constructed, he adds a new one whose abscissa is equal to the arithmetic mean of the previous ones, and whose ordinate is equal to their geometric mean.

As such, on the segment $[\alpha, 0]$, he can place all the points of the abscissas $\beta\lambda\dfrac{p}{2^n}$ (for all n and all p such that $0 < p \leq n$) whose ordinate is $\beta\lambda^{\frac{p}{2^n}}$. One thus gets a countable infinity of points, all situated on the curve of equation $y = \beta\lambda^{\frac{x}{\beta}}$, that is to say, points of a certain 'logarithmic'. Leibniz clearly recognizes this conclusion and rightly claims it.

He then proceeds to an (implicit) extension of the method, from $[0, \beta]$ to IR – the method is simple[40]. In conclusion, he has actually constructed, as he announced, "as many points as we want" of the exponential, since his method allowed him to obtain a countable infinity of points, but it does not allow him to capture an arbitrary point of it, that is to say, the *current point*. Asserting that he has constructed the exponential by means of the ordinary geometry is thus an exaggerated claim. He has nevertheless constructed an *everywhere dense infinity* of points[41] on the graphical representation of it – and this is important. One can legitimately think that Leibniz, directed by his thought of continuity, intuitively understood (and tacitly assumed) that, the set of all the constructed points being *everywhere dense*, they were allowed to capture *by continuity* an arbitrary point of the exponential curve.

The parameter of the figure is a transcendent quantity

Leibniz so constructed with ruler and compass a family of points of the curve $x \to \lambda^{\frac{x}{\beta}}$. Remained however the calculation of the value of λ. In the letter to Huygens, Leibniz considers this determination as a decisive factor (GM II, 97) because, he says, it is the ratio "constant and perpetual, valid for all the catenary lines and all their points" (96–97). He gives some explanation a little further on about this quantity, which is *the parameter of the figure* (see Figure 23):

> I have not explained yet what should be the proportion (...) from WZ to OA; but you will judge easily, Sir, that AO must be equal to the subtangent (as you call it) of the logarihmic, and that consequently, putting OW = AO, the ratio of AO to WZ is always the same and determined. So, all the logarithmic curves as well as all the catenaries, are similar or of the same species (GM II, 97).

40 To place the point W of abscissa 2β, for example, we write $A = Z^* W$; its ordinate T thus verifies $\sqrt{T\beta} = \alpha$; therefore $T = \dfrac{\alpha^2}{\beta} = \beta\lambda^2$, which actually has the announced form.

41 That is to say, given an *arbitrary point M* of the curve and an arbitrary number $r > 0$, there exists a 'Leibnizian' point L of the curve such that the distance from L to M is smaller than r.

This passage is rather elliptical; here is what is, in my opinion, its interpretation. First, Leibniz legitimately considers the only ratio of OA to WZ (i.e., two ordinates on the curve). As for OA, he had already designated it in the article as a kind of unity (GM V, 245). In the letter, he refers to the sub-tangent of "the logarithmic," that is to say, for us, a function of the form $x \rightarrow e^{k \cdot x}$. With respect to the axis Ox, the sub-tangent of this curve is constant and, for us, equal to $-\frac{1}{k}$. Therefore $\alpha = -\frac{1}{k}$. However, it is possible (see *Chapter XII, Part 2, Mathematical Study*) that Leibniz has taken a formula with an opposite value for the subtangent, and has therefore chosen $\alpha = \frac{1}{k}$ (there are two possible definitions of the subtangent, with opposite signs). As for β – of which Leibniz says nothing – it is possibly the ordinate of point of the abscissa $\left(-\frac{1}{k}\right)$ of the exponential in question. We then have $\beta = e^{-1}$ and this value is independent of k. Thus, regardless of k and whatever the exponential (or, therefore, the catenary), the value of β is directly connected to the number e, the base of the Neperian logarithms (even if Leibniz does not clarify this value). Under such conditions, one obtains: $\lambda = \dfrac{\alpha}{\beta} = \dfrac{-\frac{1}{k}}{e^{-1}} = -\dfrac{e}{k}$.

Accordingly, for Leibniz, the knowledge of the value of e is necessary and sufficient to construct the exponential curve, and hence, secondarily, the catenary. It is thus this quantity – Leibniz immediately supposes it to be transcendent – which is the key, for him, in the construction of what he calls "as many points as we want" of both curves by ordinary geometrical means[42].

The catenary, starting from the logarithmic

Once the construction of the logarithmic is granted, that of the catenary follows easily by ordinary geometric means. We explain this in Figure 23, above, from the letter to Huygens, which is clearer than that of the article. Leibniz in particular uses a relevant symmetry symbolism: A and (A) designate two points symmetrical with respect to the vertical axis.

Indeed, it is sufficient on the curve of the equation $y = e^x$ to take two points ζ and (ζ) of opposed abscissas, that is (- t) and (t). Their ordinates on the exponential are thus $N\zeta = e^{-t}$ and $(N) (\zeta) = e^t$, of which we must take the half-sum (the arithmetic mean), which is made geometrically via the point B of the figure on the axis OA; this in turn provides the point (C) on the vertical line $(N) (\zeta)$: (C) is the corresponding point of the catenary.

---

42   This conclusion is reflected in [Parmentier 2004], 188, n. 11.

The logarithmic, starting from the catenary

Leibniz then deals with the reciprocal problem (i. e., conversely), namely to construct the exponential knowing the catenary. For this, it is sufficient in the equation $f(x) = y = \text{ch } x$ to take for all $a > 0$, $x(a) = \ln a$, then $y(a) = \dfrac{a + \dfrac{1}{a}}{2}$. If the value of a is known, the quantity $y(a)$ can be constructed from a by means – again arithmetical-geometrical – of ruler and compass (and one has $y(a) \geq 1$). Considering that this value is the ordinate of a point on the catenary, one determines (graphically) its abscissa, which is $\ln a$[43]. The catenary thus enables us to construct $\ln a$ if we know a. As such, if the catenary can be entirely constructed by points – the same is true of the curve $x \rightarrow \ln x$ – and therefore (by reciprocity) of the curve defined by $x \rightarrow e^x$. As the catenary has a mechanical construction by means of a thread, this thus allows Leibniz, by a *substantially physical process*, to exhibit the value of $\ln a$ when one knows a[44].

Leibniz then proposes a comprehensive inventory of the differential properties of the catenary, that is to say, the determinations of tangents, arc lengths, quadrature, the centre of gravity of an arc, or the centre of gravity of a plane sector, etc. The catenary is indeed an exemplary curve for us – as for Leibniz – because *all its standard differential elements* of the first-order (those that Leibniz cites) as well as of the second-order can be easily determined from its equation by taking the abscissa for a parameter. For a curvilinear abscissa s, we have, for example, in modern terms $s(x) = a \, sh\dfrac{x}{a}$[45]; and for the radius of curvature R (it is a differential element of the second order) $R(x) = a \, ch^2 \dfrac{x}{a}$[46]. The catenary thus admits the intrinsic equation[47]

$$a \cdot R = s^2 + a^2.$$

43 For any number $y(a) \geq 1$ given, there are two solutions in x with respect to a, which are opposite, namely $\ln a$, and $-\ln a = \ln \dfrac{1}{a}$. We decide then the solution according to whether $a > 1$ or $a < 1$.

44 Leibniz also demonstrates how, conversely, given $\ln a$, one can determine a (GM V, 245). See [Keller 2009], 7.

45 One actually has (see for example [Serfati 1995], 168–169):

$$\left(\frac{ds}{dx}\right)^2 = \frac{(dx)^2 + (dy)^2}{(dx)^2} = 1 + \left(\frac{dy}{dx}\right)^2 = 1 + \left(sh^2 \frac{x}{a}\right) = ch^2 \frac{x}{a}.$$

Therefore $\dfrac{ds}{dx} = ch\dfrac{x}{a}$ and $s = a \, sh\dfrac{x}{a} + s_0$ (the orientation of the curve is determined by increasing values of the abscissa).

46 $R = \dfrac{\left(1 + \left(\dfrac{dy}{dx}\right)^2\right)^{\frac{3}{2}}}{\left(\dfrac{d^2 y}{dx^2}\right)}.$

47 See [Serfati 1995], 169.

"The construction of Leibniz is the most geometrical possible ..."

A new article of Leibniz devoted to the catenary, once again published in the *Acta*, in September 1691 (GM V, 255–258), provides interesting details. This is *De solutionibus problematis catenarii vel funicularis in Actis Junii an. 1691, aliisque a dn. Jac. Bernouillio propositi*s ('On the solutions of the problem of the catenary, or funicular curve, proposed among others by James Bernoulli in the *Acta* of June 1691'). He first gives a new historic presentation of the catenary. Next, after presenting considerations about the osculating circle at its current point, Leibniz reaffirms the excellence of his mode of construction – point by point – of this curve:

> For me, I finally return everything within the logarithms and, in this way, I get *the most perfect mode of expression and – at the same time – construction of the transcendent curves* (GM V, 253).

A third article (GM V, 263–266), contemporaneous to the previous article (1692), reflects this especially rich period of reflections on the subject from Leibniz. It still deals with the issue of pointwise constructions. The article, *Solutio Illustris problematis a Galilaeo primum propositi de figura, chordae aut catenae e duobus extremis pendentis, pro specimine novae analyseos cjrca infinitum* ('Solution of the famous problem proposed for the first time by Galileo, that of the rope or catenary suspended between two extreme points, as an example of the new analysis of the infinite') was published in 1692, in the Giornale di "Litterati dell" an. 1692. Modena, pp. 128–132. Leibniz first criticizes, once again, the position of Descartes (see below a note about this) and then, speaking of himself in the third person, he reaffirms the excellence of the principle of his pointwise construction:

> (...) the problems and the lines transcending Algebra (those that Descartes had been forced to exclude from his geometry because they did not fall under his calculation) (...). Yet, the construction of Leibniz is the most geometrical possible, and we cannot give another which is of a better kind, because a certain proportion being given once and for all, we can find a very large number of points of the required curve, that is to say, as much as we want, by the ordinary geometry, without assuming the quadratures, which is, in algebra, the best possible approach. That is why I have the pleasure of explaining it to a certain number of people (GM V, 263–4).

To find points "as much as we want?"
Or, "any point?"

Thus, Leibniz once again made claims here regarding the construction of "a large number of points of the required curve," that is, he says, "as much as we want" but not its current point. Such questioning will ground our epistemological reflection upon the two possible concepts for the construction of a curve 'by points' (see below).

The fourth, and last, article which we will now examine also belongs to the same period and develops similar conclusions, this time on the basis of the construction of a transcendent curve different from the catenary, namely the *Paracentric Isochrone*. This takes place in the second part of the *Constructio Propria*

(…)[48], which appeared in the *Acta* 1694 (GM V, 311–312). Leibniz is back, with more strength still, on his construction by points and making a very clear declaration of the principles encompassing both his current and previous works on the issue of how to construct a transcendent curve by giving as many points as we want of it, by means of the only ordinary quantities, and by introducing a single 'extraordinary' constant, that is to say, a transcendent one:

> When, previously, I examined the diverse methods for constructing the transcendent curves, I brought to light the most general case, namely the one by which we can determine as many points as we want using only ordinary quantities (i. e., algebraic) by supposing, nevertheless, that it is given a unique transcendent constant, valid for all the points. This is because, otherwise, it would be an eternal task to provide, for each point, new transcendent constants. This is the method that I used for constructing the catenary (GM V, 311).

Among the transcendent curves, the 'percurrent' curves are a *minima*

We will now return to another important point highlighted by Leibniz: in the above constructions, either that of the catenary – be it point by point or material (i. e., by means of a thread) – or else by that of the paracentric isochrone, none requires quadratures. This is a merit of the point by point constructions, and it is another important reason why Leibniz became attached to their study. Indeed, he willingly insists upon this aspect (GM V, 244, for example).

Leibniz called *percurrent* such transcendent curves – that is to say, those that can be constructed geometrically – point by point – by ordinary constructions, using only one transcendent quantity. The paradigmatic example, for him, would remain the catenary (and with it, 'the logarithmic'). Over time, the term 'percurrent' would, however, become a source of misunderstanding between himself and John Bernoulli (see *Section V-E*, in this chapter).

Leibniz clearly favours those percurrent curves in his analysis as being ***minimal transcendent curves***: certainly, they are not Cartesian – since they are not algebraic – but their transcendence does not require any mechanical means such as a thread nor any purely mathematical one, as quadratures. Of the latter, Leibniz knows that they can mechanically produce transcendence (see *below, Chapter VI: Quadratures, the inverse method of tangents and transcendence*), but without the geometer can really master the modalities of his action; in other words, for Leibniz, the quadratures opened up to transcendence an area that was certainly very broad –

---

48 The full title is: *Constructio propria problematis de curva isochrona paracentrica, ubi et generaliora quaedam de natura et calculo differentiali osculorum, et de constructione linearum transcendentium, una maxime geometrica, altera mechanica quidem, sed generalissima. Accessit modus reddendi inventiones transcendentium linearum universales, ut quemvis casum comprehendant, et transeant per punctum datum.* That is to say: 'Construction of the problem of the Paracentric Isochrone curve, with more general studies on the nature and the differential calculus of osculations, as well as on the construction of the transcendent curves, the one perfectly geometrical, the other mechanical, but nonetheless the most general possible. We add universal methods for determining the transcendent curves, so that they include all cases and that they pass through a given point'.

but perhaps it was too broad, because it was insufficiently delimited. Instead, the transcendence of percurrent curves is situated and summarized in a single given number (i. e., a transcendent one), for example, the base of the natural logarithms e. The percurrent curves thus represented, for Leibniz, one more milestone in his strategy of the inventorying of the non-Cartesian field.

Descartes and the 'pointwise' constructions

The criticism of Descartes by Leibniz – of which we quoted a few lines above – refers to Descartes's views on the legitimacy of the construction of curves by points, which he proposes in his *Géométrie*. For him, this legitimacy was coextensive with one of his acceptability criteria for curves in geometry. We know that there were two such criteria for Descartes[49]. I will focus, here, only on the second, the *algebraic* criterion of the acceptability of *geometric* curves (this is connected with the possibility of constructing, point by point, the roots of the equation). In a previous article[50], I have pointed out how it had been a criterion of reason governed by a matter of secondary effectiveness, and which appears completely only in 1637. From this perspective, however, the quadratrix is rejected by Descartes in the field of the 'mechanical' curves as not being amenable to the construction of *any* of its points. Descartes indicates:

> (…) In the latter not any point of the required curve may be found at pleasure, but only such points as can be determined by a process simpler than that required for the composition of the curve. (AT VI, 411).

On page 411, in the margin and under the title "Quelles sont les lignes courbes qu'on décrit en trouvant plusieurs de leurs points qui peuvent être reçues en Géométrie" ('Geometric curves that can be described by finding a number of their points'), Descartes returns to the issue of the legitimacy of the determination of a curve by "an infinity of points," which ended his previous paragraph. He distinguishes, once again (and continues to reject), the "spiral and those [that are] similar," for the reason that we cannot find any point of it, that is to say, a point of arbitrary abscissa, but those only which "can be determined by a process simpler than that required for the composition of the curve."

Descartes reports that, for some curves, such as the quadratrix or the spiral, one can determine with complete accuracy certain particular points, such as those of the abscissa $1/2$ or $1/2^n$, or even $p/2^n$ (this was noted by J. Vuillemin), but not the standard point in the truest sense, that is to say, the point of an arbitrary abscissa. Accordingly, Descartes writes about the spiral:

> Therefore, strictly speaking, we do not find any one of its points, that is, not any one of those which are so peculiarly points of this curve that they cannot be found except by means of it (AT VI, 411).

---

49   See our detailed study in [Serfati 2011b].
50   See [Serfati 2011b]. This section of the chapter widely draws upon the conclusions of this text.

In contrast, the fact that the abscissa (for example) of the point of a curve can be arbitrarily chosen – without reservation – is, for Descartes, ontologically essential to the constitution of the curve as such, that is, ultimately, to its individuation. Moreover, this is what makes so successful the Cartesian definition of geometrical curves:

> On the other hand, there is no point on these curves which supplies a solution for the proposed problem that cannot be determined by the method I have given (*idem*).

The mathematician as a constructor or as a prescriber?

As we have seen, Leibniz was satisfied to construct a very large number of points on the curve, though without asking for more. We also saw that Descartes demonstrated a quite different type of requirement. Leibniz claimed to the construction of "a very large number of points of the required curve," that is, he says, "as much as we want," but not its current point. This leads us to naturally distinguish between the two possible concepts of the construction of a curve 'by points'. For Leibniz, it is sufficient to construct 'as many points as the constructor wants'. The constructor is thus at the same time the prescriber (such a position can be associated with the synthesis). Another conception (of which Descartes was undoubtedly the champion) would be to 'construct the current point' of the curve, i. e., 'any point' of it. In other words, for a working mathematician, this is to construct a point of which the prescription may come from outside himself. The constructor and the prescriber are not here necessarily the same person.

V-E PERCURRENCE AND TRANSCENDENCE, LEIBNIZ AND JOHN BERNOULLI – THE STORY OF A MISUNDERSTANDING

The catenary, regarded as an avatar of the strategy of a geometrical inventory

Equipped with this concept of percurrence, Leibniz returned to the fundamental question in his strategy of the analysis of the non-Cartesian field, namely: can we, with this concept alone, inventory the unknown domain? In other words, are all the transcendent curves also percurrent? The fundamental nature of this question, for Leibniz, initially allows us to understand the tenacity that he deployed in relation to the study of the catenary over the years 1690–1694. A positive answer to the question would have indeed supplied a definitive end to the mystery of the non-Cartesian field – geometrically, at least. Because – and this is the second lesson of the introduction of this questioning – Leibniz, who had spent so much time on the study of particular transcendent curves – with letteralized exponentials – had also turned away from them at a certain time – he probably resigned himself to accept that they would not be able to fully explain all the transcendence that had appeared in the curves. Without giving up on the issue, he thus pushed to one side the symbolic criterion embodied in letteralization and turned instead towards the geometric criterion related to the definition 'by points' of the transcendent curves.

This aspect is rarely mentioned by commentators (with the notable exception of Breger, who analyses it in the aforementioned article (p. 129)). He cites, in this regard[51], two later items of correspondence of Leibniz to John Bernoulli, from July and October 1695. First, that of July 29, 1695, in which Leibniz wrote:

> I believe we need to think about whether all transcendent curves are not at the same time percurrent; it is possible, because the question has never been explored (…). And this is the first of my wishes and I recommend it to those of you among the bravest and the most perceptive. (A III, 6B, N.154, p. 467)

Leibniz and Bernoulli – the birth of a misunderstanding

In a letter to Bernoulli dated 10/30 October of the same year, 1695, Leibniz continues in the same vein in his search for a "peak of the geometry of the transcendence." As such, he continues talking about curves:

> We cannot say definitively that all the transcendent curves are at the same time percurrent, as you call them, that is to say describable by separate points, in accordance with the ordinary geometry. It is, however, that I suspect it to be true of the largest number, but I still do not see what distinguishes them from others. I confess that it would be the peak of the geometry of the transcendent if the thing was brought to be proved, as I remember that you said it at another time. However, we have not shown it yet (…) (A III, 6B, N.169, p. 523)

This letter reflects the continuity of the objectives of Leibniz (namely, inventorying the transcendent) as well as his pragmatism and modesty: he is not sure, he writes, that his hypothesis is true because he cannot yet see the exact location of the demarcation line between percurrent curves and transcendent curves. Yet it also reflects a misunderstanding; for when Leibniz mentions to John Bernoulli "percurrent curves, as you call them," he is mistaken. Certainly, he himself denominates as 'percurrent' those curves constructible by points. However, as we shall see, what John Bernoulli calls *percurrentes*, for his part, are curves whose equations are equipped with letteralized exponentials, which are precisely those that Leibniz had already highlighted in detail, and indeed for quite a long time previously (see above)!

Bernoulli and the percurrent calculus

This discussion between our two protagonists thus initially led to a misunderstanding. To illustrate this, we shall return to one-year previously and John Bernoulli's letter to Leibniz dated May 1694, which sheds light on the differences in their conceptions. Bernoulli reveals to Leibniz the origins of his 'percurrent Calculus':

> These equations differ from your transcendent ones in that, in yours, the number of dimensions is vague and indeterminate, in mine, determined but unknown. Since percurrent curves of this style demand a particular system in order to determine their points, tangents, quadratures, etc., I already have, for a long time, coined the idea of a percurrent Calculation where I deal with

---

51   [Breger 1986],129, n. 63 et n. 64.

how to sum the differentials of the percurrent equations and construct all the percurrent curves by means of the Logarithmic vulgar, which, as I conceive it, is itself a percurrent curve; its equation, indeed, if we call x the abscissa, y the ordinate (*applicata*), and a the sub-tangent, is $a^x = y$ (…) (A III, 6A, N. 35, p. 91).

Bernoulli then takes the example (p. 140) of the percurrent curve of equation $y = x^x$. This correspondence, sent to Leibniz in 1694, may have appeared naive to the latter! This is because, for a long time and as we saw it in detail, Leibniz himself had carefully studied the equations of curves equipped with letteralized exponentials, that is to say, the 'percurrent Calculation' in the sense of Bernoulli, even if he had not given it this name. Moreover, Leibniz had also wanted to take into account the distinction – contained in the first sentence quoted from Bernoulli – between equations of indefinite degree and equations of unknown degree (however, as we also saw above (*Chapter IV-D2*), this distinction is somewhat illusory).

What John Bernoulli was searching for had thus been examined sometime previously by Leibniz. Bernoulli simply brought a new terminology; he called *percurrent* the equations which contain a letteralized exponential, and *percurrent curves*, the curves equipped with such equations – they are specific transcendent curves – the initial example being the logarithmic. He so speaks about a *percurrent Calculation*, even trying to establish a hierarchy in percurrence which he thus defined (see *infra*).

A middle way between algebraicity and transcendence?

Bernoulli would resume this argument shortly afterwards, in two new letters to Leibniz from 1695. First, from 17 July 1695:

> I have already enough shown to the Marquess of the Hospital where Craig had made a mistake, but I do not think we should make it public before Craig himself has replied to my objections. Beyond these transcendent curves which I mentioned to you in my last correspondences, and which are certainly a part of those whose points we can have algebraically, I see that they all belong to the domain of those whose nature can be expressed by an equation, the dimension of which goes so far as the indeterminate, as $x^x = y$. From these, we can even reach the quadratrix, the Archimedean spiral, the rhumb curve and other plane curves. Indeed, it is possible, even in these curves, to determine any of their points. I leave to your reflection as to whether I had, or not, rightly called the curves of this kind of the particular name 'percurrent', by distinguishing them from those generally that are transcendent, that is, those of which you cannot find an arbitrary point in an algebraic way; and whether they can be, or not, considered a middle way between algebraic and transcendent curves (A III, 6B, N. 150, p. 456–457).

The objective of Bernoulli is clear ("finding a middle way between algebraic and transcendent curves"), but his preamble is somewhat problematic: where Leibniz had carefully spoken of determining "as many points as we want" of the logarithmic, Bernoulli asks much more by asserting that it is possible to determine "any point" of it. The difference is not small, since the latter process is generally impossible, even by giving the number e as an auxiliary transcendent constant (see the previous *Section V-D* of this chapter); this is an additional and crucial aspect which Bernoulli, unlike Leibniz, does not even mention. Two months later, Bernoulli wrote to Leibniz again on the subject, from Basel, on August 27 (September 3)

1695, to confess "I have difficulty in thinking that all the transcendent lines are at
same time percurrent" (A III, 6B, N. 157, p. 484). Later, in 1695 (17/27 December)
he sent a new letter to Leibniz on the same issue, from Groningen. Bernoulli was
quite clear at this time, both as to the precise definition that he gives of the percurrent curves and as to his own epistemological issues:

> I strongly doubt whether it is a peak of the geometry that to know if the transcendent curves can
> be reduced to percurrent, that is to say, to curves the equations of which consist of terms with
> indeterminate dimensions. This is why, even now, I am of the opinion that all the percurrent
> curves can be constructed by means of the quadrature of the hyperbola: but in reality, you take
> them with a wider meaning of the word. Indeed, for me, the Quadratrix of the circle is not per-
> current, because its nature cannot be expressed by such an equation (A III, 6B, N. 189, p. 587).

In this correspondence, Bernoulli says, in essence, that the percurrent equation $y =
a^x$ leads to:

$$\ln y = x \ln a = kx \text{ and therefore to } x = \frac{1}{\ln a} \ln y$$

We thus obtain the curve in the question by means of the "logarithmic" – the latter
arising from the quadrature of the hyperbola. However, and rightly says Bernoulli,
not all quadratures reduce to that of an hyperbola. For the squaring of the circle,
Bernoulli is indeed right. As we saw it in detail in *Chapter II*, we need an additional
argument for it – a symbolic substitution – that is foreign to the quadrature of the
hyperbola, namely the replacement of x by x^2, in the rational fraction of the equation $y = \frac{1}{1+x}$.

The end of an ambiguity – the 'hypo-transcendence' of the exponentials

The previous letter clearly explains not only the conceptions of John Bernoulli but
also his differences in terminology with Leibniz. It is indeed clear that, for Bernoulli, the only definite percurrent curves are those curves with letteralized exponentials, which Leibniz had considered much earlier. When he writes to Leibniz
that the latter takes the term 'percurrent' curves "with a wider meaning of the word,"
he obviously refers to the design that Leibniz had conceived (in the years 1690–91)
of "curves constructed by points" (cf. previous section). The ambiguity of the terminology is thus revealed at exactly this moment.

   This conclusion is widely confirmed by an important essay which John Bernoulli would publish two years later, in the *Acta* of March, 1697, entitled *Principia
Calculi Exponentialum, seu percurrentium*[52]. This is a very interesting report, completely dedicated to what Bernoulli called, from then on, "Exponential Calculation,
that is to say, Percurrent." After a few lines for Nieuwentijt (p. 179), he commences
(p. 180) a clear presentation – sometimes dogmatic – on percurrence and exponentials, and which is valid for both equations and curves:

---

52   *Oeuvres*. Bd. 1, p. 180 N° XXXVI.

I had called 'percurrent' a quantity raised to an indeterminate power (before I had realized that it had already been considered by Leibniz), because it runs through all the possible dimensions. It is for this reason that I have designated by the same name both the equations made of quantities of this kind and the curves that they define. But, since this naming 'exponential' is a pleasure to our great mathematician, in his honour, I take for myself this other name; I say this first, so that – what often happens – different names for the same thing do not raise confusion with the reader.

So, I designed the exponential quantity as a middle way between the algebraic and the transcendent: it is close to the algebraic, since it is made of finite terms, i. e., indeterminate; but, for the transcendent, we cannot exhibit any algebraic construction (*Œuvres* I, 180).

We thus understand better the vanity of the earlier controversies between Leibniz and Bernoulli, which were based on a misunderstanding. In fact, the latter had taken ownership of the equations and the curves with letteralized exponentials, which Leibniz had previously sought to describe with so much patience and care, before moving away from them because they were, for him, of an inadequate extension (and thus insufficiently 'explanatory' of transcendence). For Leibniz, in fact, the exponential symbolism could not describe the whole of geometric transcendence. On the other hand, Bernoulli, who had given them the label 'percurrent', was capable of considering a 'percurrent calculation'. For his part, after his initial symbolic exponential period, Leibniz refocused – as regards curves – on construction 'by points' (see the previous *Section V-D*) as the criterion of selection – and these curves he called *percurrent*.

On a possible hierarchy in percurrence

Furthermore, in the aforesaid report, Bernoulli examined the organization of degrees in percurrence via the repetition of exponentiation:

The exponentials can present different degrees; the smallest (the only one so far examined) is recognized when the exponential is composed of ordinary undetermined quantities, such as y^m, x^n and x^p, where m, n and p are simply undetermined quantities. An exponential quantity is said to be of the second-degree when the exponent itself is an undetermined quantity, as in y^{m^n}; and so on. In a general way, an exponential quantity of some degree has for the exponent an exponential quantity of an immediately lower degree (*Works* I, 182).

In this elaboration by induction of a hierarchy in percurrence, we recognize in Bernoulli an expression of these reasoned inventories, glorified so much by Leibniz (see above, *Section V-C*). Furthermore, Bernoulli calculates tangents (by differentiation) and areas (by quadratures) for various models of his percurrent curves (see Examples III and V below) (p. 186):

$$x^x = a^y \text{ (Example III)}$$

$$x^x + x^c = x^y + y \text{ (Example V)}$$

We propose, here, a brief study in modern terms of Bernoulli's solution for Example V. He wants to differentiate $x^x + x^c = x^y + y$ with respect to x and y, that is to say, to calculate a total differential. One has $x^x = e^{x \ln x}$, and therefore:

$$d(x^x) = (1 + \ln x) x^x dx \text{ and } d(x^c) = cx^{c-1} dx.$$

Similarly, $d(x^y) = yx^{y-1} dx + x^y \ln x \, dy$ (total differential). Note that the theorem on the differential of a composite function appears here only in an implicit form. We thus have, finally:

$$(1 + \ln x) x^x dx + cx^{c-1} dx = yx^{y-1} dx + x^y \ln x \, dy + dy$$

V-F MATHEMATICAL COMPLEMENTS – ENVELOPES AND EVOLUTES

Determination of the envelope of a family of curves

I expose below a very brief modern presentation of the theory of envelopes in the plane[53].

Theorem (envelope of a family of plane curves). Let C_λ ($\lambda \in I$) (I is an interval of IR) be a one-parameter family of plane curves, with differential class C^1; its equation is:

$$F(x, y, \lambda) = 0 \ (1)$$

The envelope (E) of the family is obtained by the system (S), consisting of (1) and relation (2) below:

$$\frac{\partial F}{\partial \lambda}(x, y, \lambda) = 0$$

In other words, ***one derived with respect to the parameter*** λ, by considering x and y as constant: this embodies the aforesaid *exchange of interpretations*.

In modern terms, the standard previous Leibnizian procedure had consisted of considering λ to be fixed and differentiating the equation with regard to x (x is thus variable; λ is arbitrary but fixed): this was the context of the ordinary differential calculus. As such Leibniz found the tangent at the point of abscissa x to the curve of parameter λ.

Leibniz now inverts the points of view, that is to say, he differentiates the same equation this time with regard to λ (x and y are now arbitrary but fixed); accordingly, he obtains the equation of the envelope of the family. Relation (2) can be considered as the equation of another one parameter-family of curves, namely H_λ: for each fixed λ, the intersection $C_\lambda \cap H_\lambda$, if it exists, thus provides the characteristic point of the envelope.

To solve (S), we can either solve it with regard to x and y as functions of λ – we thus obtain a parameterization of the envelope (E) – or else eliminate λ between (1) and (2); in the algebraic case, we can also write that (1) admits a double root with regard to λ . Of course, it is possible that none of these eventualities are actually achievable.

---

53   See [Serfati 1995], Chapter VI, *Propriétés métriques des arcs*, 167–212.

Application: The envelope of a family D_λ of straight lines of the plane is a simple special case of the above. Let us take for F the equation:

$$F(x, y, \lambda) = u(\lambda)\, x + v(\lambda)\, y + w(\lambda) = 0 \ (1)$$

where u, v and w are C^1-functions on I and $(u(\lambda), v(\lambda)) \neq (0,0)$.
    Relation (2) is therefore:

$$u'(\lambda)\, x + v'(\lambda)\, y + w'(\lambda) = 0 \ (2)$$

Geometrically, this involves the intersection $D_\lambda \cap D'_\lambda$ of *two one-parameter families of straight lines*. We thus know how to solve it with regard to x and y, and so find a parameterization of the envelope in the neighborhood of any point where $u(\lambda)\, v'(\lambda) - u'(\lambda)\, v(\lambda) \neq 0$.

Evolutes and involutes

Theorem and definition. The locus (D) of all the centres of curvature[54] among the diverse points of a plane arc (C) – of a differentiability class C^2 – coincides with the envelope of the normals to (C). We call this the *evolute* of (C), which is, in turn, an *involute* of (D).

This result is fundamental in differential geometry. It expresses that every normal at a point M of a curve (C) 'touches' its envelope (D) at a point which is precisely the centre of the osculating circle in M to (C).
    A plane arc being given, it thus admits exactly one evolute. On the other hand, the reciprocal problem – that is, the determination of the arcs of which it is the evolute – has (in general) infinitely many solutions. Here, we detail this point.
    Let $(\delta) = (I, \overline{F})$ be a parameterized arc of class C^1 on which one chooses a curvilinear abscissa $u \to s(u)$.[55] For every real number a, the parameterized arc $(\gamma_a) = (I, \overline{G}_a)$ where

$$\overline{G}_a(u) = \overline{F}(u) + (-s + a)\, \overline{t},$$

is an involute of (δ).

$$(\overline{t} = \frac{d\overline{F}}{ds} \text{ is the unitary tangent vector to } (\delta)).$$

Alternatively $\overrightarrow{OM} = \overrightarrow{OC} + (-s + a)\, \overline{t}$. There is thus an infinite number of involutes of (δ).

---

54   That is to say, the centres of the osculating circles.
55    $\overline{F}, \overline{t}$ etc. denote vectorial functions.

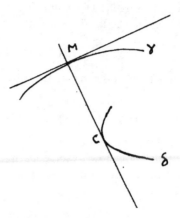

Curves without equations – physical constructions

The aspects of curves as material objects appear very early in Leibniz, since, as already noted above, one of the earliest instances of the term – in the draft of the 1675 letter to Oldenburg – already refers to the curves so obtained in the context of the typology of quadratures. Leibniz considers, among other quadratures, the one that he calls a 'mechanical quadrature':

> (…) which can be exact, but is performed using certain material curves provided with a certain extent, as by the unwinding of a thread (A III 1, 202).

Let us repeat the point: this is an early text, at a time when Leibniz had only just discovered the issues of quadratures and had not yet succeeded in a more assertive reflection on symbolism. It is, however, necessary to note that, even in his years of maturity, Leibniz would never give up the practice of material curves and threads for the purpose of constructing curves when he could not do otherwise; the process allowed him to solve the case of some quadratures. The paradigmatic example is, obviously, the catenary, as we saw it above in *Section V-D*.

CHAPTER VI
QUADRATURES, THE INVERSE METHOD OF TANGENTS, AND TRANSCENDENCE

QUADRATURES AS A 'BLIND' MODE OF GENERATING TRANSCENDENCE

The issue of quadratures will remain, throughout Leibniz's work, a major mathematical question. It first appears as a double-sided process, on the one hand symbolic, on the other hand geometrical, before it becomes, from the 1690s onwards, a somewhat 'blind' and mechanical, privileged source for generating transcendence. I detail here an example given by Breger[1], in *De Geometria recondita et Analysi Indivisibilium atque Infinitorum* (=*On the Hidden Geometry and the Analysis of Indivisible and Infinite Quantities*), published in the *Acta* of June 1686 (GM V, 226–233). This well-known text is of great interest regarding the relationship between quadratures and transcendence. In his introduction to the French translation, Parmentier rightly notes that, for Leibniz, the question in this text is to place:

> the new calculus in its most natural and most fruitful perspective, namely the evaluation of quadratures, and more specifically, the inverse problem of tangents, of which the quadratures are but a special case[2].

Leibniz initially quotes Craig, before presenting a long history of the issue, and writes:

> To come to more useful things, it now seems appropriate to *uncover the source of transcendent quantities*[3] and to reveal why some problems are neither plane, nor solid, nor sursolid, nor of any determined degree, but they exceed[4] any algebraic equation. By the same way, we shall show how we can prove, without calculations, that it is impossible that there exists an algebraic quadratrix of the circle and of the hyperbola (GM V, 228).

As we see, Leibniz is here fully in line with his permanent strategy of inventory. He indeed proposes to "uncover the source of transcendent quantities", *i. e.*, to elucidate – at last – the whole mystery of non-Cartesian knowledge. Further, he observes:

> However, the very method of investigating indefinite quadratures, or their impossibilities, is for me only a particular (and indeed simpler) case of a much wider problem, which I call the *inverse method of tangents* in which is contained the largest part of all transcendent geometry.

---

1   [Breger 1986],127, and notes 48, 49.
2   [Parmentier 1989], 127.
3   Emphasis in the text.
4   Leibniz uses the Latin verb *transcendere*.

Leibniz, here, rightly distinguishes between quadratures and the inverse tangent method, that is to say, in modern terms, respectively, between integration and the resolution of differential equations (see below). Breger summarizes and confirms this point of view[5]:

> As we repeatedly said, transcendence in Leibniz is mostly related to problems of quadrature. In the essay from *Acta Eruditorum* (1686), in which appears the first sign for the integral in a printed text, Leibniz aims "to uncover the source of transcendent quantities," and he explicitly stresses that the largest part of the geometry of transcendence is being developed within the inverse method of tangents. It is not surprising: tangents to algebraic curves can be calculated in an algebraic way, but it happens only to a limited extent for the quadratures of such algebraic curves. Difficulties with respect to the formulation of the integral calculus, occurring because of this fact, were the main motivation for Leibniz to devote himself in a global way to the problems of transcendence.

Recall, incidentally, the place – analogue as well as dissymmetrical – that Leibniz assigned to the calculation of evolutes, with respect to this function of systematically generating transcendence (see above V-C). I shall also underline that, in 1730, John Bernoulli still naively continued to conceive transcendent functions as necessarily associated with quadratures[6]: for him, being given an (almost) arbitrary function, one systematically generates transcendence by – 'blindly' – putting just before it the sign of integration $\int$[7].

The typology of quadratures in Leibniz

Let us now turn the clock back, that is, to Leibniz's initial representations of various types of quadratures. As we have seen, squaring the circle was, from 1673 onwards, his great subject, and the matter of a very important theorem. The questions of quadratures naturally were of great importance for him, which they remained until the end of his life.

During the period 1675–1680, he was repeatedly engaged in attempts relating to the inventory and typology of quadratures, as we have already discussed in Chapter I. What might surprise the modern mathematician reader is probably the mixing of genres: where we now see a major topic of mathematical analysis (i. e., integration), we observe that the important issue in Leibniz's eyes was the ability to square, whatever the way. So, Leibniz accepts ***mechanical quadratures***, through curves or cables (see above, V-C), and also ***geometrical quadratures***, by means of 'ordinary geometry' (see, for example, the catenary in Chapter V-D).

However, he distinguishes from these two previous kinds the ***analytical quadratures***, which, he said, are the only ones through which "calculation can be done" (GM V, 120). He still subdivides this 'analytical' section into three sub-sections: algebraic, arithmetical, and transcendent. *De Vera Proportione* (…) of the *Acta*,

---

5   [Breger 1986],127 and notes 48, 49.
6   John Bernoulli, 'Méthode pour trouver des Tautochrones, dans des milieux résistants, comme le carré des vitesses' = *Œuvres*, vol. 3, p. 174.
7   This is the object of Breger's remark ([Breger 1986], 128 and n. 56).

1682 (GM V, 119–120), proposes a perfectly structured development on classification and subdivisions.

An ***analytical algebraic*** quadrature fits with the very simple idea that, from an algebraic equation, one can algebraically calculate it. For example, the quadrature of the parabola of equation $y = x^2$ gives $A(x) = \dfrac{x^3}{3}$ for the area between the vertex and the point of abscissa x. For a 'paraboloid' of equation $y = x^n$ (n is a natural integer), one obtains $A(x) = \dfrac{x^{n+1}}{n+1}$ for the area between the points of abscissa 0 and x.

In contrast, the idea of an ***arithmetical analytical*** quadrature was certainly grounded, for Leibniz, on his result about the arithmetical quadrature of the circle, which he obtained using Mercator's method, with the addition of the term-by-term integration of some power series (see above, II-A). In addition to the introduction of power series, based on squaring paraboloids (see above), was the – absolutely extraordinary for the time – use of an infinite sum. Let us suppose, in modern terms, the initial curve has as an equation $y(x) = \displaystyle\sum_{n\geq0} a_n x^n$; then its quadratrices are given by

$$A(x) = \sum_{n\geq0} a_n \frac{x^n}{n+1} + C,$$ C being an integration constant. The integer (n+1) necessarily appears in the denominator. The adjective 'arithmetical' probably comes from this example, even if its terminology now appears somewhat naïve (it did not survive Leibniz). Probably, Leibniz stores under this 'arithmetical' section all the quadratures obtained through term-by-term integrated power series.

The final step in the classification, namely the idea of a ***transcendent analytical*** quadrature, certainly appears later. Its origin was the failure of explicit results – even in the case of curves with very simple algebraic equations, the circle and the hyperbola remaining basic examples. Breger's comment, quoted earlier in this chapter, is quite relevant.

As we have seen, Leibniz first tries to convince himself that every curve, even if it is not algebraic, and if nevertheless it admits an equation, the latter can be written with letteralized[8] exponentials, so that one can express the quadrature of the curve. Letteralized exponentials would then be the top of the theory of quadratures. This conception was embodied in what I called the 'exponential utopia', above, which Leibniz exposed in many texts, especially in the abovementioned letter to Tschirnhaus, May 1678. Breger stresses[9]:

> A few years before these publications (1678) in the *Acta Eruditorum*, he [Leibniz] advises Tschirnhaus of the difficulties of the integral calculus (GM IV, 456–458). To find all the quadratures, or to prove the impossibility of an (algebraic) quadrature, I do not see any other possibility – Leibniz writes – than solving equations in which the variable also enters the exponent (…). With the completion of this kind of transcendent calculus, everything would be solved of what is desirable as to the problems of quadrature.

---

8   As indicated above, this term is my own neologism. It characterizes the systematic substitution of a letter instead of a digit.

9   [Breger 1986], 127.

Facing the – anachronistically predictable – failure of such an attempt, Leibniz did not give up in any way, and questioned the very essence of the calculus (see below, Chap. VIII).

The inverse method of tangents

One can legitimately question the origin of the (substantial) conceptual importance of quadratures for Leibniz. Why search so hard for them? Why use them as the driving force of his analysis? Admittedly, in his early mathematical stages – with the arithmetic quadrature of the circle as a 'headlight' – he naturally favoured this approach. Admittedly, also, this issue of quadratures was wholly unfamiliar to Descartes, an aspect that could not obviously fail to reinforce Leibniz's interest. But these explanations may be somewhat limited.

Another substantive reason for the interest of Leibniz in this question certainly lies in the fundamental place occupied by the calculus of quadratures within the inverse method of tangents, briefly mentioned above. As regard to the (differential) first order, this method applies to the resolution of some problems. There follows an example phrased in modern terms:

> Find one (or several) curve(s) (C), defined by $x \rightarrow y = f(x)$, where f is a function of the differentiability class C^1, such that their sub-tangent satisfies a given 'metric' property (i. e., concerning lengths and distances), for example, where the sum of the sub-tangent and of the cube of the abscissa is constant.

Curves subject to a condition and differential equations

The inverse method of tangents is the search for one (or more) curve(s) ruled by a condition upon its (or their) tangents. Thus, the second of De Beaune's problems (see above, Chap. IV-B) pertains to this questioning. In modern terms, this issue – supposedly well carried out – leads to a first order differential equation, which must then be solved.

Generally speaking, the search for all 'conditioned' curves by what we call 'an arbitrary differential condition of the first order', leads to the formation of such a differential equation[10]. On the one hand, however, explicitly solving a differential equation is not obvious; there are many equations, even of the first order, for which no method exists for solving them as such – that is to say, as an equation. Even today, one must only develop methods for approximated resolutions.

On the other hand – this point is important – even if one can solve the equation, the calculus, inevitably as well as structurally, will end with an integration, (*i. e.,* with a quadrature); this is even the genuine pursued aim, namely to reduce the reso-

---

10   Before De Beaune and Leibniz, mathematicians (such as the Greeks, as well as Descartes) did not raise this kind of problem. Of course, they were looking for curves, which were also *conditioned*, but this condition only concerned 'metrical' aspects (i. e., dealing with lengths) and so it did not involve any differential element.

lution of any differential equation to an integration, such as: $\dfrac{dy}{dx} = F(x)$ [11]. This is a second problem, epistemologically distinct from the previous one, to which it is simply posterior. Thus, in France, the epistemological difference between both difficulties has long been explained to students by saying that a differential equation is 'reduced to quadratures', if it is solved *as a differential equation*; the question of being able – or not – to calculate the inevitable final primitives is considered to be subordinate.

One can easily understand why, for Leibniz, the inverse method of tangents has always been consubstantially linked to transcendence and really endowed with an organically higher complexity, because it ultimately involves quadratures (whereas the reverse is not true). Thus, one can better appreciate Leibniz's position in the abovementioned extract of *De Geometria recondita* (1686).

Arc lengths and rectifications

To further emphasize the importance of quadratures in the late seventeenth century, let us remind ourselves that other issues will be added to Leibniz; they certainly were distinct from the inverse method of tangents, which nevertheless ended with quadratures. Just like *rectifications*, which undoubtedly represented a class of very important problems: to rectify a curve is to determine a formula giving the length of any of its arcs. For us today, this requires the previous definitions of the concepts of arc length and curvilinear abscissa; in the seventeenth century, it was a new problem, and certainly not a simple question. I do not develop this point here[12].

We might note, in modern terms, that the rectification of a curve (C) given by $x \rightarrow y = f(x)$, where f is a function of the differentiability class C^1, leads to the quadrature[13]:

$$\int_0^x \sqrt{1 + \left(f'(t) \right)^2}\, dt$$

In respect of an inevitable radical, the rectification often leads to problems of quadrature (i. e., of integration) structurally more complicated than those of ordinary quadratures, even for simple algebraic curves. This is also the case for variational problems, such as the brachystochrone or the determination of geodesics. The resolution (1696) by Leibniz of the brachystochrone problem is remarkable, and Leibniz is fully aware[14] that he has not just solved a problem but discovered a general method applicable in many other cases of variational problems, such as that of geodesics posed by Jean Bernoulli (see A III, 7 N. 149 N. 215).

---

11   To do so, usually we use a change of variable and / or function.
12   See, for example, [Hofmann 1974], 101–117, *The quarrel over rectification.*
13   For a modern presentation, see [Serfati 1995], 167.
14   See A III, 7 N. 29.

On 'constants of integration'

To clarify, I will finally return to an important aspect exposed above in IV-D5. For Leibniz, it has been important to understand that a plane curve can be known – just as well, and sometimes better – by a differential equation (for him, mostly a differential equation of the first order) than by its original equation. Such awareness, radically new in his time, is obviously a direct consequence of the conception he always made of his differential calculus. In later texts, Leibniz will, however, refine his position, by underlining a natural difference between, on the one hand, a curve given by an explicit equation, and on the other hand *via* a differential equation. In this case, the curve is not completely determined, because there will inevitably be a constant of integration. Indeed, solving a differential equation of the first order must always end with a quadrature; there will thus never be a unique solution, and we always obtain a family of solutions dependent upon a parameter, namely this constant of integration. Leibniz pertinently analyses this fact in a text from *Acta* (1703), *Continuatio Analyseos Quadraturae Rationalium* (trans. *Continuation of the analysis of the quadratures of the rational functions* (GM V, 361).

    A supplementary issue is the development of a place for this constant in the finished symbolism. Certainly, this place is clearly *additive* (with regard to y) in the terminal quadrature. The equation $\frac{dy}{dx} = F(x)$ actually leads to $y(x) = G(x) + C$, where C is the constant of integration[15] and G is a primitive function of F. But regarding the differential equation itself, the constant may ultimately appear in more surprising places, even in simple equations (such as linear or homogeneous ones) of the first order. A simple example is the resolution (for $x > 0$) of the simple homogeneous equation:

$$(y-x)+(x+y)\frac{dy}{dx} = 0$$

The resolution leads to a sheaf of hyperbolas, with the equation:

$$y^2+2xy-x^2 = C^2 \text{ (restricted to } x > 0)$$

where C, which is the constant of integration, does not appear additively, with respect to y.

---

15   x here is supposed to describe an interval I of IR. Should it describe several disjoint intervals, there is a constant for each interval. This observation fits into the more general problem of the *connectedness* of the domain of definition of the functions in question (any interval is connected).

CHAPTER VII
LEIBNIZ AND TRANSCENDENT NUMBERS

A VALUE, WHICH 'IS NOT A UNIQUE NUMBER'

The occurrences of Leibniz's reflection about transcendent numbers appear much scarcer compared with the abundance of his works on curves and expressions. One of them concerns Gregory's method for demonstrating the transcendence of π in a modern sense – of which Leibniz was aware since 1673 (see Chapter II). Leibniz took an active interest in Gregory's work, especially as regards his attempt to generalize the convergence method to the geometric-harmonic mean; but he did not develop a specific point – now of some interest – namely an undeniable characterization of the algebraic numbers (in the modern sense – see *infra* Chapter XV).

At the same time, the importance of his theorem for the arithmetical quadrature of the circle led Leibniz to focus on the nature of his own result, which provides the area limited by a quarter of a circle with radius 1 (in modern terms $\frac{\pi}{4}$). It is the sum of the numerical series:

$$1 - \frac{1}{3} + \frac{1}{5} - \frac{1}{7} + \text{ etc.}$$

Leibniz's first public description of the subject takes place in the *Acta* of 1682, with his famous article *De Vera Proportione* (...)[1]. In this well-known text – a founding document of the differential calculus – Leibniz first indicates that the resulting value is *exactly* expressed by the sum of the series. It is not an *approximation*, he says, even if *it contains all approximations*[2]. He nevertheless clearly adds that *this value cannot be a number*, just like the irrational entities cannot be numbers. Leibniz's comment to the status of his own result seems somewhat surprising to modern readers; it was, indeed, fully in line with the conceptions of his days. Leibniz writes:

> Thus, the series simultaneously contains – in its entirety – all approximations, that is to say the immediate higher values and the lower ones: indeed, the error will be less than any given fraction, and hence than any given quantity, depending on the farther extension of the series. Thus, the series in its entirety expresses the exact value. Although a single number cannot express the sum of such a series and although the series continues to infinity, however, since a unique law of progression governs it, then, the mind can perceive it in its entirety. Indeed, as the circle is not commensurable with the square, and therefore cannot be expressed by a single number, one should rather express it using rational numbers through a series, in the same way as the diagonal of a square (...) and many other irrational magnitudes (GM V, 120).

---

1   GM V, 118 –122.
2   One has, for example $1 - \frac{1}{3} < \frac{\pi}{4} < 1 - \frac{1}{3} + \frac{1}{5}$ and also $1 - \frac{1}{3} + \frac{1}{5} - \frac{1}{7} < \frac{\pi}{4} < 1 - \frac{1}{3} + \frac{1}{5} - \frac{1}{7} + \frac{1}{9}$, etc.,
      etc. This is a standard property of convergent alternating series. See [Parmentier 2004], 327.

Thus, in Leibniz's conception, the sum of the series is not really consubstantial with an individualized *number*, rather with a *process*. However, the latter is harmonious and easy to conceive ("the mind can perceive it in its entirety"). Concerning the same article, Breger pertinently comments:

> In his essay dated 1682 on the arithmetical quadrature of the circle, Leibniz confirms that the series $1 - \frac{1}{3} + \frac{1}{5} - \frac{1}{7} +$ etc. is not a simple approximation for the area of the quarter of circle with radius 1, but it is the exact expression of the value (...). However, Leibniz confesses that there is no number for this value. Wallis for example named π as "numerus ille impossibilis"[3]. To explain this, Leibniz adds that the same design applies to irrational expressions of roots and also to the relation between the diagonal and the side of a square: again, here is no number.

As such, during this initial phase, Leibniz hardly considered the specific issue of transcendent numbers, probably in order to be consistent with the prevailing opinion of the leading mathematicians of his time.

THE NUMBER IS "HOMOGENEOUS TO THE UNIT"

We can summarize the young Leibniz's initial position at the time of the arithmetical quadrature of the circle: the result of his own calculation was not a number. Over time, however, he returned – to modify them – to his youthful concepts in some theoretical manuscripts. Breger pertinently explains this aspect. At first, he notes the inflection of Leibniz's position in a 1682 text: *Initia Mathematica,* subsection *De Quantitate*[4]. Leibniz thus began to modify his concept on the nature of numbers[5] soon after the publication of *De Vera Proportione*.

He initially proposes (p. 30) a – relatively extensive – definition of equal, similar or homogeneous figures. He states:

> Similar things are such as one cannot distinguish them, if they are only considered each one individually. This is the case of two spheres, or two circles (or two cubes, or two squares) A and B (...). Homogeneous things are those which are either similar or can be made similar by a transformation.

It is the case, Leibniz says, of two straight lines, but also of a straight line and of an arc of a circle. This point is important. Indeed, if a curved line can be transformed into a straight line, then both lines are homogeneous. In modern terms, it allows so in Leibniz to consider π as a number.

Later (p. 31), he considers as definitive the definition of a number as something homogeneous to a unit:

---

3   Incidentally, note that Wallis's reference to *numerus impossibilis* – here evoked by Breger – can also be found in Huygens' work during the same period, with exactly the same words. See Huygens' manuscript dated 1686 or 1687, 'Du livre de Wallis, Historia Algebrae Anglicé. Développement du "Numerus impossibilis" (π) en une fraction continue', in [Huygens 1888], vol. 20, Chapter IV, 388–394.

4   GM VII, 31–32, dated 1682 (see Ritter).

5   In Gerhardt edition, the text begins on page 29 and follows *Initia Rerum Mathematica*.

> The number is homogeneous to the unit and, a fortiori, can be compared to the unit, added to it or subtracted (...)

For Leibniz, from then on, the following entities are thus numbers: integers, fractions, $\sqrt{2}$ (namely an irrational or surd number), or still better, as he says, *ineffable* (p. 31) – the diagonal of the square is indeed homogeneous to the side. This consideration seems sufficient to Leibniz, and relieves him from the necessity of any reasoned additional inventory.

THE TRANSCENDENT NUMBERS IN *DE ORTU*

Another theoretical text, dated 1685, *De Ortu, Progressu et Natura Algebrae, Nonnullisque Aliorum et propriis Circa Eam Inventis*[6] (i. e., 'On the origin, development, and nature of Algebra and some discoveries about its subject') clarifies the matter. In this text, Leibniz explicitly speaks of *transcendent numbers*. However, these appear without definition, simply within the framework of a long list – a typology – of various sets of opposites. Let us repeat: Leibniz, breaking with his initial conceptions, henceforth considers transcendent numbers as given numbers, but without characterizing them. In this text, Leibniz's conception is entirely devoted to reductions (p. 208): fractions can be reduced to integers, surd numbers to rational ones, 'affected' surd numbers to 'pure' ones; 'imaginary numbers' can be expressed by 'possible numbers', and finally, transcendent numbers by 'usual' ones. However, Leibniz writes that this latter reduction "is achieved by means of series" (GM VII, 208). One still finds here the resonance of his Arithmetical Quadrature of the Circle; however, with some conceptual advance: henceforth, transcendent *quantities* became *numbers*.

TRANSCENDENT NUMBERS IN THE *NOUVEAUX ESSAIS*

Finally, it is worth noting Leibniz's position – one can consider it as definitive – in Theophilus's answer to Philalethes, in the *Nouveaux Essais* (1704), Book II, Chapter XVI (*Du nombre*[7]), § 4:

> That applies to integers only. For number in the broad sense – encompassing fractions, surds, transcendent numbers and everything that lies between two integers – is proportional to a line; we do not find there any more minimum than in the continuous.

As such, the 'transcendent quantities' permanently became fully-fledged *numbers*. But their description still clearly lacks some means to capture them in a reasoned way! For the modern reader, the inventory of all kinds of numbers is nevertheless satisfactory by the image of the real line IR which it proposes. However, Leibniz

---

6   GM VII, 203–216. Ritter dated this text 1685, and Hofmann 1685–1686. We here pursue our study by using many references provided on the subject by [Breger 1986], page 131 and n. 76.
7   GP V, 142.

did not try, here, to develop a reasoned inventory, as he did for curves; for example, he did not look for a positive characterization of the numbers *that are not transcendent* (the algebraic numbers).

TRANSCENDENT NUMBERS IN INITIA
RERUM MATHEMATICARUM METAPHYSICA

A later theoretical text (1715), also quoted by Breger (note 76), *Initia Rerum Mathematicarum Metaphysica* (GM VII, 17–29) shows[8] that this conception of transcendent numbers became definitive in Leibniz. This text displays an inventory of all kinds of numbers, in the style of *Nouveaux Essais*, without any further definition of transcendent numbers. Leibniz writes:

> From this it clearly appears that the number can be generally defined, according to its kind, as integer, fractional, rational, surd, ordinal, transcendent; because it is either an entity homogeneous to the unit, or which is equivalent to a unit, just like a straight line is to a straight line (GM, VII, 24).

Thus, Leibniz resumes his argument of homogeneity, which again seems to satisfy him and to dispense him from a more accurate characterization of various types of numbers. *Initia Rerum* concludes (p. 25) on the law of continuity and on the *termini inclusivi* and *exclusivi*: each species ultimately ends with the opposite species[9].

DIRECT EXPRESSION OF A NUMBER AND EXPRESSIONS
WHICH *CONTAIN* IT

It is thus clear that, contrary to what he had quite extensively done for curves and expressions, Leibniz considered transcendent numbers in a slightly more superficial way. Before Leibniz, Gregory was certainly the one who came closest to the issue. After Leibniz, Lambert would definitively formulate the question under a modern form (see Chap. XIII-B).

Gregory was concerned with demonstrating that π could not be expressed as an algebraic function of known (rational) quantities. Moreover, we saw that Leibniz, in his brief commentary on the transcendent numbers, had concluded in a similar way that π, as it was expressed with an infinite series (in the Arithmetical Quadrature of the Circle), could no more be expressed as an algebraic number.

For both Gregory and Leibniz, it was thus an (implicit) philosophy of the *direct expression* of a number that was operating there; such a conception seems quite natural. Lambert's merit was, however, to upset it, by considering not the *status of the direct expression* of a number but *the status of a quantity which contains it* – namely, a polynomial with integer coefficients that *contains* algebraic numbers as its roots.

---

8   Ritter dates the text April 1715.
9   On this point, see [Serfati 2010a], 12.

THIRD PART
TOWARDS AN "ANALYSIS OF THE TRANSCENDENT"

INTRODUCTION

The third part of this volume consists in two chapters, VIII and IX, respectively entitled "On Harmony in Leibniz" and "From the Harmonic Triangle to the Calculus of Transcendents".

As we saw, Leibniz's efforts to inventory the transcendent in geometry came up against the extreme diversity of the geometrical objects under consideration. He certainly discovered new practices to build some of them, but he failed to promote a universal method. Leibniz was pleased in examining in detail some of the – diverse – geometrical constructions, but when he realized the limitations of these procedures, he did not insist upon them.

With letteralized exponentials, Leibniz, envisaging under another perspective the inventory, came back to equations, expressions and symbolism: a matter with which he also felt comfortable. Introducing letteralized exponentials was, at least initially, a new idea, a promising project, on the one hand deeply rooted in the post-Cartesian symbolism, and on the other hand in his own differential equations (see Chap. IV-D). He dedicated to his exponentials much of his energy and hopes – it was the time of the exponential utopia and John Bernoulli would later follow him in this way. Unfortunately, as we can well-understand it now, it was in vain: not all transcendent expressions – in the 'non-Cartesian' sense – can be described by exponential expressions – far from it. As I have already noted, Leibniz, who could have distanced himself from the issue of the inventory, certainly did not give up; on the contrary, he developed instead of the ordinary analysis an entirely new theory of calculation and symbolism, which he called 'transcendent analysis' (or 'calculation'). We will detail this in the following two chapters.

Let us repeat: Leibniz was facing a difficult and ill-defined problem – he had to inventory an entire, unknown world. Mathematically speaking, the solution that he chose came from fields different from those he had previously considered (namely, geometry and functions): these were sequences, algebra, and the structure of both triangles, arithmetical and harmonic. From a methodological viewpoint, the solution he had chosen arose from a reasoned practice of *harmony*, in the sense that he gave to this term in his philosophy.

The conclusion was as follows (and we develop it in this part of the volume): Leibniz's discovery of a common structure to both aforesaid triangles, despite the fact that they were structurally opposed, and *in fine,* because of this very opposition, led to the construction – in the calculation – of a harmony through reciprocity, that of both assemblers of signs d and ∫. In our modern mathematician opinion, this approach did not (and could not) bring a definitive solution to the issue of the inventory; however, it symbolically built, and in a definitive way, a new scheme of calculation grounded on the differential and integral calculus.

Leibniz's approach towards what he later called "New Calculation of Transcendent" culminated, in 1694, with the publication in the *Journal des Sçavans* of an article of the utmost importance, *Considérations sur la Différence qu'il y a entre l'Analyse Ordinaire et le Nouveau Calcul des Transcendantes*. We analyse this text below in IX-D).

CHAPTER VIII
ON HARMONY IN LEIBNIZ

VIII-A *UNITAS IN VARIETATE*

Harmony is all the greater as it is revealed in a greater diversity

It is well known that the doctrine of harmony is central in the philosophy of Leibniz. The subject has resulted in a rich philosophical commentary[1]. Here is, for example, Yvon Belaval:

> The Leibnizianism appears as a monadological entity that we can examine from many 'points of view', under many perspectives, and this harmonic structure alone would testify that the idea of harmony is the fundamental insight ([Bélaval 1976], 86).

We shall not detail here all the inflections of meaning of Leibniz on this point, but propose further a quite explicit Leibnizian text. The central point is that harmony is recognized and characterized, says Leibniz, as "unity in diversity". In these conditions, it is interesting to point out some variations in his terminology. I resume here another quotation from Bélaval:

> Leibniz often defines harmony as *unitas in varietate* (…). Instead of *unitas*, we find *simplicitas* (…), *similitudo* (…), *identitas* (…), *consensus*, as instead of *varietas*, we find *multa* (…), *plura* (…), *multitudo* (…), *diversitas* (…). To the quantitative theme of the one and the multiple, gets married the qualitative theme of the same and the miscellaneous (*idem*, 87–88).

This unity (or identity) and this diversity (or still this variety) can be thus recognized in the qualitative domain as well as in the quantitative. From this central doctrine, Leibniz derives diverse conclusions – which seem to me to be as important as necessary. He acknowledges, for example, that there are degrees in harmony, and that harmony is all the greater as it is revealed in a greater diversity. Even the dissonances contribute to the creation of harmony, provided they are reintegrated within the framework of 'unity'[2]. So understood, the 'requirements' of harmony, the search for the latter, and its uncovering, constituted for Leibniz a regulative principle of thought and action[3].

The 'perfection of cogitability'

Among the many texts on the subject, I have chosen, in the *Elementa verae pietatis* of 1677–78[4], what Leibniz calls a 'corollary' in which he clearly explains how, in

---

1 See, for example, [Bélaval 1976] and [De Buzon 1995].
2 See our study on this point in [Serfati 2013a].
3 See 'L'harmonie comme norme', in [De Buzon 1995], 110.
4 A VI, 4, N. 256 = Grua, 12, 108–109.

his view, harmony is the perfection of the faculty of thinking (*perfectionem cogita-bilitatis*):

> Harmony is when many things are reduced to some unity. For where there is no variety, there is no harmony, and we laugh at the Citharede (…)[5] but you know the continuation. Conversely, where variety is without order, without proportion, without accordance, there is no harmony. Hence, the greater the variety and the unity in variety, the greater the harmony. From there, dissonances themselves increase accordance if soon they are reduced to concord with other dissonances. The same is true with symmetries. Hence it easily comes that harmony is the perfection of cogitability (…) It follows that this mode of thinking is the more perfect, that is to say the one by which a single act of thought extends to many objects at the same time: so there is more reality in that thought. This is done by means of relations; for a relation is indeed a certain unity within multiplicity (…) Of all these relations simultaneously enrolled in a given object results a harmony. So, the more relations there are in a thinkable object (the aggregate of which is harmony), the more there will be reality or, what is identical, perfection in the thought. As a result, harmony is the perfection of thinkable things, to the extent obviously that they are thinkable.

Harmony and the identity of indiscernibles

Breger, placing himself within the framework of the physics in Leibniz and especially the harmony coextensive with a possible symmetry of the objects, explains clearly how[6] in Leibniz's system of the world, the universal harmony principle is a necessary counterpart to that of the identity of indiscernibles (and, as we know, the latter is also one of the fundamental principles of the metaphysics of Leibniz). Indeed, if this principle provides a natural richness and variety of a world of objects, it could simultaneously produce an unintelligible world, because of this same diversity and the possibility of 'unlimited differences' (according to Breger).

This conclusion concerned the state of the world. On the other hand, however, the human mind must be organized so that it can fully grasp the underlying common structure to a diversity of objects. In addition, the thought is thus all the more harmonious (i. e., for Leibniz, more perfect) that it is able to capture the structure under a greater diversity:

> According to Leibniz, the thought is all the more perfect as at every stage several objects are simultaneously involved. What he calls the fundamental rule of his philosophical system is that things that arise with unlimited differences must be understood by the same basic principles. Leibniz sees the beauty and the perfection of the universe in the interplay between, on the one hand the wealth of things, and on the other the identity of structures. ([Breger 1988], 26)

Accordingly, seeking harmony is equivalent to uncovering the underlying structure to an 'unconstituted aggregate'[7].

---

5   This is a reference to Horace's well known verse in the Poetic Art (verses 335–337): "They make fun of a Citharede which is always wrong on the same string."
6   [Breger 1988].
7   My remark refers to an "unconstituted aggregate of disunited people," that is to say the famous phrase describing, according to Mirabeau, the state of France before the Revolution.

Harmony in the mathematics of Leibniz

Leibniz did not limit his conception of harmony to only philosophical contexts. Instead, he willingly claimed the practice of harmony in various mathematical situations. In the (symbolic) context of the consistency of a linear system of equations, elimination and determinants, he so writes to Hospital: "We find everywhere harmonies which serve us as guarantees"[8].

In a previous article [Serfati 2013a], I discussed some mathematical practices of harmony in Leibniz, trying to answer two questions. First, how can we recognize (and then analyse) the practice of harmony in Leibniz's mathematics? Secondly, what can we say, historically, of its usefulness for the advancement of mathematics? In the same text, I also recalled that before undertaking any study, it was first necessary to stress this fundamental methodological observation: Leibniz's practice of harmony took place within the framework of symbolic writing; this activity was still new at that time, but had however already spread through all mathematics[9]. I shall only sketch the general argument that I have developed at greater length and more accurately in another paper.

The mathematics of Leibniz teem with examples in which he looks at and reveals a certain unity in the symbolic text, even while he is confronted with the diversity of some structures of mathematical writing. To search for and to bring to light this unity is exactly what his practice involves. From this point of view, he was followed by posterity in a direct way, and to the most contemporary mathematics. Nonetheless, what we meet here at first is thus diversity, that is, the variety of the mathematical symbolic representations, and the research consists in discovering some permanence: in other words, what is 'the same' (or sometimes, only, 'a same')? 'The same' is thus an embodiment of a structure, and it is always by proceeding in this direction that Leibniz initially discovers harmony, that is to say: he goes from *Varietas* to *Unitate*, and not the reverse.

Symbolism and social communication

Unity had thus to cover, in a given context, all the occurrences of the symbolism in question. On the other hand, one of the issues raised in the late seventeenth century by the extensive use of the new symbolism was certainly the need to find a way to communicate the results of the mathematical research which was, from now on (and definitively!), symbolically written[10]. Because the symbolism brought not only advantages and privileges, it also conveyed some new constraints, which were undoubtedly *social*. And, this was also a problem to which Leibniz became attached extensively. We can thus see how the search for harmony allowed him, at first, to

---

8   Letter to L'Hôpital, dated 28th April 1693 (A III, 5B, N. 148, p. 546).
9   On the philosophical and epistemological aspects of the introduction of symbolism, see in our book [Serfati 2005] the sections 'Caractéristique et "Nouveau Calcul" chez Leibniz', pp. 269–284 and 'L'Art Combinatoire', pp. 285–321.
10 See [Serfati 2001] and [Serfati 2006a].

bring to light the concept of some unity, and then to express this new concept through the symbolism by means of a representation immediately communicable to the general public.

Harmony *via* reciprocity, *via* homogeneity, *via* symmetry

Let us return more concretely to mathematics, in Leibniz, by distinguishing the diverse fields for implementing harmony, among which three, namely harmony via reciprocity, harmony via homogeneity, and harmony via symmetry, are particularly important. Note that the latter two instances, symmetry and homogeneity, are directly connected with algebra rather than with analysis. Both were coextensive in Leibniz with the use of specific symbolizations, firstly his 'fictitious numbers'[11], then his diacritical 'two-dots'[12] – it was the first symbol of operational exhaustiveness in the history of mathematics. I detailed precisely these algebraic aspects in other preliminary texts. In the present work dedicated to transcendence, I shall limit myself to the study of harmony by reciprocity in Leibniz's calculus and analysis.

VIII-B THE TWO TRIANGLES AND THE LETTER TO L'HOSPITAL OF DECEMBER 1694

I shall begin by describing the history and some of the properties of the harmonic triangle; they are often contemporary to Leibniz, while some others are modern. The properties of this triangle testify to an exceptional mathematical inventiveness on the part of Leibniz. We shall therefore have to deal with an epistemological scheme common to both triangles, namely: arithmetical (Pascal) and harmonic (Leibniz). This harmony led Leibniz to the idea of a certain structure of calculation.

During the year 1672, in the context of his discovery of the differential calculus, Leibniz invented the harmonic triangle. For him, this structure always possessed an outstanding mathematical as well as rhetorical value, and he considered it to be an essential part in the development of his thought. On the other hand, the harmonic triangle remained, until today, a mathematical specificity of Leibniz (sometimes called 'Leibniz's triangle').

We shall examine, first, a letter of December 27th, 1694, of Leibniz to L'Hôpital. It is a famous correspondence in which, in particular, Leibniz confesses: "My metaphysics is all mathematics, so to speak, or it can become so" (AIII, 6A, N.84, p. 253). On the other hand, the text admirably describes the diverse stages of the rise and development of the Harmonic Triangle:

> I had taken pleasure, long before, to look for the sums of series of numbers; and I had made use for this purpose of the differences, via a well-known theorem, namely that if a series is decreas-

---

11   See 'Élimination, nombres feints et situs', in [Serfati 2001], pp. 168–173.
12   For example

$$10.21.30.$$

ing at infinity, its first term is equal to the sum of all the differences. This gave me what I called the Harmonic Triangle, opposed to Pascal's Arithmetical Triangle. Because Pascal had shown how we can give the sums of the figurate numbers, which arise by looking for the sums, and also for the sums of sums of the terms of the natural arithmetic progression; and I found that the fractions of the figurate numbers are the differences of the terms of the natural harmonic progression, that is to say the fractions

$$\frac{1}{1}, \frac{1}{2}, \frac{1}{3}, \frac{1}{4}, \dots,$$

and that we thus give the sums of the series of the figurate fractions as

$$\frac{1}{1} + \frac{1}{3} + \frac{1}{6} + \frac{1}{10} + \dots$$

$$\frac{1}{1} + \frac{1}{4} + \frac{1}{10} + \frac{1}{20} + \dots$$

So, as I recognized this great utility of the differences and as I saw that, by Descartes's calculation, the ordinate of the curve can be expressed, I saw that finding the quadratures, or sums of ordinates, is nothing else than to find an ordinate (of the quadratrix), the difference of which is proportional to the given ordinate. I realized also soon that finding the tangents is not something else than to differentiate, and that finding the quadratures is nothing else than to sum, provided we assume the differences to be infinitely small (…) And that is the story of the origin of my method (AIII, 6A, N.84, p. 256).

In this text, Leibniz rightly sums up the diverse stages of the creation of the Harmonic Triangle. It was first coined *in opposition to* the arithmetic triangle; then, it would become a central piece in the method of the 'sum of all the differences'. Afterwards, it became a major mathematical instrument used by Leibniz to calculate some sums of numerical series with positive terms. Later still, it gave rise to the creation of a computational structure equipped with two entirely new antagonist Leibnizian operations, symbolized by d and ∫. Finally, it served as a support to ensure the permanence of the validity of these properties in geometry (at least, Leibniz hoped for this!). We detail these various steps below.

VIII-C PASCAL AND LEIBNIZ.
THE FIRST CONSTRUCTION OF THE TRIANGLES (BY *ELEMENTS*)

As can be understood from the above extract, one of the origins of the harmonic triangle was Blaise Pascal's arithmetic triangle. It had been well understood, very early on, by Leibniz. Early in his Paris period, Leibniz had indeed plunged into Pascal's diverse works, to which he would make many references. An interesting article of Costabel[13] examines this question. For example, in his Arithmetical Quadrature of the Circle, Leibniz made major use of the *Traité du Sinus de Quart de Cercle*[14]. However, he also made systematic use of the *Traité du Triangle Arithmétique*.

---

13 [Costabel 1962].
14 Cf. Costabel, op. cit., 'Triangle caractéristique et calcul différentiel', 369–371.

A section (pp. 371–373) of the article quoted from Costabel is moreover dedicated to 'Arithmetical Triangle and Mathematical Analysis'.

There were, in Pascal, two versions of the arithmetical triangle; the first one in Latin, dated 1654, was not spread. The second, in French and Latin, was published in 1665[15]. In Paris, Leibniz discovered Pascal's triangle in this last version; he will make frequent references to it during his Parisian period (and also beyond)[16]. He also annotated the margins of his copy of the book. This copy and his notes are kept in the Leibniz Library, in Hanover, where I was able to consult them, and are dated (by S. Probst), from 1672 until 1676.

Construction 'by elements' of Pascal's arithmetical triangle[17]

$C(n, p)$ denotes the element of the n-th row and the p-th column of the arithmetical triangle. In modern terms we have:

$$C(n, p) = C(n-1, p-1) + C(n-1, p) \ (\forall \ n,p) \ (n-1 \geq p \geq 1)$$
This is the recurrenc*e* Relation (1), and

$$C(n, 0) = C(n,n) = 1 \ (\forall \ n \geq 1)$$
Source Relation (2)

Hence the elementary figure:

p-1	p		
C(n-1, p-1)	C(n-1,p)		n-1
	C(n,p)		n

---

15 See [Mesnard 1964] (vol. II – 1990). Also see a comment in [Descotes 2008], 240–241. Also [Edwards 1987].

16 Thus, AVII, 3 N 4 contains numerous references to Pascal and to the arithmetical triangle, for instance 'De Artibus resolvendi progressionem irreductam', pages 30–49. This text is dated July-December 1672 and contains four parts. See also the manuscript AVII 3, N532, (february 1676) = 'Triangulum harmonicum respondens triangulo arithmetico Pascalii, quo ostenditur quomodo numerorum figuratorum reciproci seu fracti, possint addi in summam. Patet et quomodo saepe ex simplicibus divinemus composita', pages 708–711. The latter text is quoted, with others, by Costabel (p. 371–372).

17 We explain below, in IX-A, the description of the second method of the construction of the arithmetical triangle ('by lines').

and the scheme:

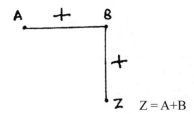

Z = A+B

Therefore, the general presentation of the Arithmetical Triangle is:

p \ n	0	1	2	3	4	5
1	1	1				
2	1	2	1			
3	1	3	3	1		
4	1	4	6	4	1	
5	1	5	10	10	5	1

Construction 'by elements' of Leibniz's harmonic triangle

The procedures of Leibniz and Mengoli (see below, Chapter IX-C) upset Pascal's design – in two separate directions – by introducing two algebraic reciprocal schemes, namely, differences instead of sums for the recurrence relation, and reciprocal integers instead of integers for the source relation (i. e., the generating sequence). One thus has:

$$K(n, p) = K(n-1, p-1) - K(n, p-1) \ (\forall \, n, p) \ (n-1 \geq p \geq 0)$$

Recurrence Relation (3), and

$$(\forall \, n \geq 0) \ K(n,0) = \frac{1}{n+1}$$

Source Relation (4)

Hence the elementary figure:

	p-1	p	
	K(n-1, p-1)		n-1
	K(n, p-1)	K(n,p)	n

and the scheme:

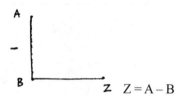

$$Z = A - B$$

Therefore, the general presentation of the harmonic triangle is:

n \ p	0	1	2	3	4	5	6
0	1						
1	$\frac{1}{2}$	$\frac{1}{2}$					
2	$\frac{1}{3}$	$\frac{1}{6}$	$\frac{1}{3}$				
3	$\frac{1}{4}$	$\frac{1}{12}$	$\frac{1}{12}$	$\frac{1}{4}$			
4	$\frac{1}{5}$	$\frac{1}{20}$	$\frac{1}{30}$	$\frac{1}{20}$	$\frac{1}{5}$		
5	$\frac{1}{6}$	$\frac{1}{30}$	$\frac{1}{60}$	$\frac{1}{60}$	$\frac{1}{30}$	$\frac{1}{6}$	
6	$\frac{1}{7}$	$\frac{1}{40}$	$\frac{1}{105}$	$\frac{1}{140}$	$\frac{1}{105}$	$\frac{1}{40}$	$\frac{1}{7}$

Based on this full structural opposition between two schemes, Leibniz brought to light the permanence of a fundamental underlying structure (see *infra*, *The Scheme* in Chapter IX-D). From there, he allowed himself to solidly develop three new figures of thought. First of all, the 'sum of all differences', then the creation of his higher-order differentials, and finally the generation of double-entry tables (starting from a simple line); all three became, under diverse aspects, fundamental elements of his 'new calculation'.

VIII-D HUYGENS, LEIBNIZ AND THE SUM OF RECIPROCAL TRIANGU-
LAR NUMBERS (*FIRST METHOD*)

The problem of Huygens

In this section, we develop some historical aspects[18]. Leibniz, starting from the arithmetical triangle, where the generating sequence is made entirely of all natural integers, gradually established the harmonic pattern to which I have just referred.

In the manuscript *Origo inventionis trianguli harmonici*, dated December 1675-February 1676 (AVII, 3 N53₃, p. 712), Leibniz traces back the idea of his invention to a conversation with Huygens dated ('probably', according to Probst[19] September 1672. Huygens asked Leibniz to determine the sum of the infinite series of the inverses of triangular numbers (in modern terms $\sum_{n\geq 0} \frac{2}{n(n+1)}$). Through a somewhat fortuitous method, Huygens had found its exact value (namely 2) many years previously (in 1665) but he had not published it[20].

Leibniz's first method

Leibniz found the same result by *two methods*: first, by a purely computational method (with, however, a mistake). The procedure is nevertheless interesting, and we briefly consider it below. It is described in detail in the *Summa Fractionum A Figuratis Per Æquationes* (AVII 3, N. 36 p. 365). Leibniz defines $B = \sum_{n\geq 1} \frac{1}{n}$ (which does not exist because the series diverges!) and $A = \sum_{n\geq 1} \frac{1}{n(n+1)}$ (i.e., the half-sum of the reciprocals of triangular numbers), and he tries to calculate 2A via a relation between A and B. He finds that B − A is equal to 'almost B'; in fact: B − A = B − 1, so that B can be removed, and that A = 1; thus 2A = 2, such that Leibniz writes:

$$\frac{2}{1} = \sum_{n\geq 1} \frac{2}{n(n+1)}$$

The method is thus incorrect, because of the divergence of the harmonic series. We could, however, now make it correct, by using partial sums instead of the sum of series.

By a very similar calculation, Leibniz finds $\frac{3}{2}$ for the sum of the reciprocal pyramidal numbers (p. 369) (i.e. $\sum_{n\geq 1} \frac{6}{n(n+1)(n+2)}$). In this case, his method is

---

18   This historical study is completed and extended below in Chapter IX.
19   See [Probst 2007].
20   See *Oeuvres* = [Huygens 1888], Tome XIV, 144–150.

perfectly correct, as it is for the calculation of the sum of the reciprocals of the figurate numbers of higher orders.

The sums of reciprocal figurate numbers.
Analysis of Leibniz's first method

In fact, the calculation of Leibniz can be expressed in modern terms as follows: define z_p as the searched sum of the reciprocals of figurate numbers of the p-th order:

$$z_p = \sum_{n\ge 1} \frac{p!}{n(n+1)...(n+p-1)}$$

We know that z_1 is divergent, and we have just seen that $z_2 = 2$. We also observe that the first term of z_p is always 1. Let us look for a recurrence relation between z_p and z_{p-1}:

$$z_{p-1} = \sum_{n\ge 1} \frac{(p-1)!}{n(n+1)...(n+p-2)}$$

To this end, Leibniz, led by his first two examples above, calculate the linear combination: $\frac{p-1}{p} z_p - z_{p-1}$

$$\frac{p-1}{p} z_p - z_{p-1} = \sum_{n\ge 1} \left[\frac{p-1}{p} \cdot \frac{p!}{n(n+1)...(n+p-1)} - \frac{(p-1)!}{n(n+1)...(n+p-2)} \right]$$

$$= \sum_{n\ge 1} \frac{(p-1)!}{n(n+1)...(n+p-2)} \left[\frac{p-1}{n+p-1} - 1 \right]$$

However, the value of the last term (in square brackets), which is (-n), is independent of p. This is the beauty of the linear combination ... Thus, by changing signs:

$$z_{p-1} - \frac{p-1}{p} z_p = \sum_{n\ge 1} \frac{n \cdot (p-1)!}{n(n+1)...(n+p-1)}$$

Dividing by n (it was also one of the aims of the operation!), it remains, with the changing of indices $q = n+1$:

$$z_{p-1} - \frac{p-1}{p} z_p = \sum_{q\ge 2} \frac{(p-1)!}{q(q+1)...(q+p-2)}$$

In other words, it is 'almost z_{p-1}' (namely, z_{p-1}, minus its first term). Hence:

$$z_{p-1} - \frac{p-1}{p} z_p = z_{p-1} - 1$$

Thus, despite appearances, the method of Leibniz is *not* a recurrence, because we have, directly:

$$z_p = \frac{p}{p-1}$$

for the value of sum of the reciprocals of figurate numbers of the p-th order (p≥2).

Particularly, $z_2 = 2$ and $z_3 = \frac{3}{2}$, but, as mentioned, the calculation of z_2 has assumed, in Leibniz, that of z_1, which does not exist. In the manuscript (AVII, 3 N533), Leibniz will thus want, by a purely symbolic analogy, to also apply it in the case of p = 1, thus writing $\frac{1}{0} - a$ result of which we would say that it reflects the divergence of the harmonic series. But this case, Leibniz writes (page 714), is obtained in a speculative manner (*conjecturalia*) unlike the other calculations, which are certain (*certa*):

A	B	C	D	etc.
$\frac{1}{1}$	$\frac{1}{1}$	$\frac{1}{1}$	$\frac{1}{1}$	$\frac{1}{1}$
$\frac{1}{1}$	$\frac{1}{2}$	$\frac{1}{3}$	$\frac{1}{4}$	$\frac{1}{5}$
$\frac{1}{1}$	$\frac{1}{3}$	$\frac{1}{6}$	$\frac{1}{10}$	$\frac{1}{15}$
$\frac{1}{1}$	$\frac{1}{4}$	$\frac{1}{10}$	$\frac{1}{20}$	$\frac{1}{35}$
$\frac{1}{1}$	$\frac{1}{5}$	$\frac{1}{15}$	$\frac{1}{35}$	$\frac{1}{70}$
etc.	etc.	etc.	etc.	etc.

	summa		

$\frac{0}{\ldots}$	$\frac{1}{0}$	$\frac{2}{1}$	$\frac{3}{2}$	$\frac{4}{3}$	etc.

coniecturalia   certa

Harmonic progressions

This was therefore the first method of Leibniz. As to the second, it will be precisely registered in a triangular construction derived from Pascal's arithmetical triangle. This was another origin of the harmonic triangle. We develop this further in Chapter IX-A.

On the other hand and again, the contemporary resolution of the Arithmetical Quadrature of the Circle, conducted by Leibniz, involved in a natural way the alternating harmonic series (it is now known to be conditionally convergent):

$$\frac{\pi}{4} = 1 - \frac{1}{2} + \frac{1}{3} - \frac{1}{4} + \text{etc.}$$

Its general term is no longer composed of integers but rather of reciprocal integers in arithmetic progression, i. e., $\frac{(-1)^n}{an+b}$; its sum contains in particular sums of terms of the form $\frac{1}{an+b}$, as in the harmonic triangle.

Leibniz called such series *harmonic progressions*. In various manuscripts, he studied their properties, for instance in AVII, 3, N27; N28; N55. He also defined them in his *Treaty on the Arithmetical Quadrature of the Conics* (see [Parmentier 2004], Proposition XXXIII, 220). When added to Pascal's arithmetical triangle as well as to Huygens' question, his Arithmetical Quadrature of the Circle thus constituted for Leibniz a third source of interest for the harmonic triangle.

VIII-E) A MATHEMATICAL COMPLEMENT: THE DETERMINATION OF THE ELEMENTS OF THE HARMONIC TRIANGLE

We proceed by induction from relations (3) and (4) above:

$$K(n, 1) = K(n-1,0) - K(n,0) = \frac{1}{n} - \frac{1}{n+1} = \frac{1}{n(n+1)}$$

In particular $K(1,1) = \frac{1}{2}$

Then

$$K(n,2) = K(n-1, 1) - K(n, 1) = \frac{1}{(n-1)n} - \frac{1}{n(n+1)} = \frac{2}{(n-1)n(n+1)}$$

$$K(n,3) = K(n-1,2) - K(n, 2) = \frac{6}{(n-2)(n-1)n(n+1)}$$

Take for the induction hypothesis:

$$K(n,p) = \frac{p!}{\left[n-(p-1)\right]\left[(n-(p-2))\right]\ldots n(n+1)} \text{ (there are p+1 factors in the denominator).}$$

Or else $K(n,p) = \frac{p!(n-p)!}{(n+1)!}$ (5) $\left(= \frac{1}{(n+1)\binom{n}{p}} \right)$

One can easily verify that $K(n, 0)$ is equal to $\frac{n!}{(n+1)!} = \frac{1}{n+1}$.

Suppose, then, that formula (5) is true at the order (p–1), for every n. Hence

$$K(n-1,p-1) - K(n, p-1) = \frac{(p-1)!\left[n-1-(p-1)\right]!}{n!} - \frac{(p-1)!\left[n-(p-1)\right]!}{(n+1)!}$$

$$\frac{(p-1)!(n-p)!}{n!}\left[1 - \frac{n-p+1}{n+1}\right] = \frac{p!(n-p)!}{(n+1)!} = K(n,p)$$

The value of K(n, p) is thus validated by induction. It follows, in particular, that K (n, p) = K (n, n-p), and therefore so does the *structural symmetry* of the harmonic triangle.

CHAPTER IX
FROM THE HARMONIC TRIANGLE
TO THE CALCULUS OF TRANSCENDENTS

IX-A THE SECOND CONSTRUCTION OF THE TRIANGLES (BY *LINES*)

The Leibnizian logical pattern

In the previous chapter, we presented the first construction of the two triangles – the most natural – by means of their elements. Leibniz realized, however – he was the first to do so – that, whatever the triangle, arithmetical or harmonic, its central mathematical object was indeed the *line* (namely, a sequence of real numbers), and not the element, as one might spontaneously imagine by referring to the initial construction of each triangle, as did (in their time) Pascal and Mengoli. This awareness was certainly the origin, in Leibniz, of a *functional* point of view of calculation, as opposed to Descartes' *numerical* point of view (see below, *Descartes and Leibniz, numbers versus functions*).

In fact, Leibniz considered that, if one moves from one line to another line of the same triangle, we proceed inevitably, either by summation or by differentiation, as appropriate. However, he also understood, at the same time, that this order should be reversed when passing from the arithmetical triangle to the harmonic triangle (see Figure 1, *infra*). Furthermore, it was a fundamental reflection by him on the double, antagonistic structure of the two triangles.

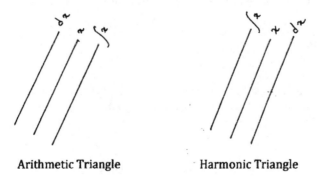

Arithmetic Triangle   Harmonic Triangle

Figure 1. The first logical pattern of the two triangles.

Leibniz developed his conclusions in numerous texts. As already noted above in a letter to de l'Hospital. He will repeat them in *Historia and Origo Calculi Differentialis*, September, 1714:

In the Arithmetical Triangle any given line is the line of the sums of the one which precedes it immediately, and the line of the differences of the one which follows it immediately. In the Harmonic Triangle on the contrary, it is the sequence of the sums of the one which follows it immediately, and the line of the differences of the one that precedes it immediately (GM V 405).

The Leibnizian pattern: presentation and justifications

We have thus just exposed the results claimed by Leibniz. We are interested now, naturally, to explain and to justify them. In order to establish his results, Leibniz started from the following statement: "the sum of all the differences is equal to the difference between the extreme terms." It was a statement that was important for him, that he even happened to place "at the origin of the mathematics"[1]. We present below a shortened modern version of his method.

To any sequence of given real numbers denoted x, with a general term x_n $(n \geq 1)$, we associate two other sequences. First, the *differences sequence* of x, denoted dx, the general term of which, $(dx)_n$, is defined by:

$$(dx)_n = x_n - x_{n-1} \ (n \geq 2) \text{ and } (dx)_1 = x_1 \ (1)$$

Next, another sequence, called the *sums sequence* of x, denoted $\int x$, with the general term $\left(\int x \right)_n$ defined for $n \geq 1$, by:

$$\left(\int x \right)_n = \sum_{1 \leq k \leq n} x_k \quad (2)$$

The latter is thus simply the partial sum of the rank n of the series with the general term x_n. From this, one can simply prove that d and $\int$ are two reciprocal one-to-one mappings on the set $S = \mathrm{IR}^{\mathrm{IN}}$ of all the sequences of real numbers.

In fact, for any sequence x (x = line of triangle), we have:

$$\left(d\left(\int x \right) \right)_n = x_n \ (3) \text{ and}$$

$$\left(\int dx \right)_n = x_n \quad (4)$$

These two formulas are fundamental: they are indeed *structurally valid* on the set of all the sequences, and do not presuppose any kind of convergence. We now present a proof.

Proofs of these results

a) $d\left(\left(\int x \right) \right)_n = \left(\int x \right)_n - \left(\int x \right)_{n-1} = \sum_{1 \leq k \leq n} x_k - \sum_{1 \leq k \leq n-1} x_k = x_n$ for $n \geq 2$.

---

1   See a letter from Leibniz to Clarke dated late 1715 = GP VII, 355–356. I detailed this point in [Serfati 2013b].

While for n=1: $\left(d\left(\int dx\right)\right)_1 = \left(\int x\right)_1 = \sum_{k=1} x_k = x_1$ (Relation 3)

b) For n= 1, we immediately have $\left(\int dx\right)_1 = \sum_{1 \le k \le 1} dx_k = dx_1 = x_1$

While for n≥2,

$$\left(\int dx\right)_n = \sum_{1 \le k \le n} (dx)_k =$$

$$(dx)_1 + [dx_2 - dx_1] + [dx_3 - dx_2] + ... + [dx_{n-1} - dx_{n-2}] + [dx_n - dx_{n-1}] = dx_n$$

This is indeed the **sum of all the differences** (Relation (4)), so that:

$$d\left(\int x\right) = x \quad \text{et} \quad \left(\int dx\right) = x$$

Hence,

$$d\left(\int \ \right) = Id_s \quad \text{and} \quad \int(d) = Id_s \text{ (Relations 5)}$$

Our terminology ('reciprocal one-to-one mappings') is obviously modern. At the time of Leibniz, the reciprocity contained in both formulas (5) was certainly interpreted as the emergence of a pair of opposites, or else that of two competing processes, the successive execution of which leads to the identity, whatever the direction, in which the operations are performed (on this point, see [Serfati 2008a], pp. 128–130).

However, the same results (5) can also be interpreted as a **dissociation of identity**: this one, which is considered as primitively given, indeed splits according to the succession of two reciprocal procedures. This dissociation is thus perfectly embodied by the method of the sum of all the differences. Moreover, this one had itself originated in the arithmetical triangle, and then in the harmonic triangle.

On convergence. The second method of Leibniz
for the reciprocal triangular numbers

The previous conclusions obviously apply whatever the nature of the sequence x in question. However, if we can additionally have the **convergence** of the sequences involved, then the method allows us, by a natural passage to the limit, to calculate easily some sums of numerical series, and in particular to find by a new method a fundamental result, the subject of Huygens' question (see above, Chapter VIII-D), namely the sum of the reciprocal triangular numbers.

Indeed, with $x_n = \dfrac{1}{n}$, we have $dx_1 = 1$ and $dx_n = -\dfrac{1}{n(n-1)}$ for n≥2. From formula (4), one then gets:

$$x_n = \left(\int dx_n\right) = \sum_{1 \le k \le n} dx_k = 1 - \sum_{2 \le k \le n} \frac{1}{k(k-1)} = x_n = \frac{1}{n}$$

Since the sequence $\dfrac{1}{n}$ converges to 0, by passing to the limit one obtains Huygens' result $\left(\displaystyle\sum_{k\geq2}\dfrac{1}{k(k-1)}=1\right)$.

Higher-order differentials

On the other hand, the simple visual inspection of the diagram of the first logical pattern (Figure 1, above) naturally suggests the introduction of higher-order differentials (denoted $d^{P}x$) as so many other possible interpretations of the lines of each triangle, according to Figure 2 below:

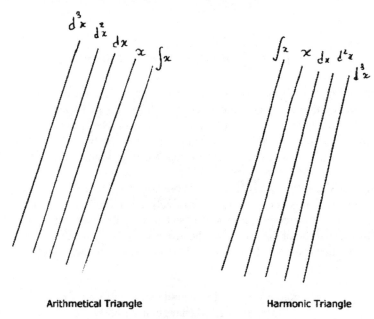

Arithmetical Triangle   **Harmonic Triangle**

Figure 2. The second logical pattern of the two triangles.

Therefore, as Granger pertinently notices[2], this introduction of a higher-order differential does not involve any infinitesimal! The use of formulas (5) applied to successive higher-order differentials $d^{P}x$ of the harmonic triangle thus allowed Leibniz to give an extended response (according to his second approach) to the question of Huygens: it included the reciprocal triangular numbers as well as the reciprocals of higher-order figurate numbers. Indeed, Leibniz could naturally apply to the *sequence* dx what he had just said of the *sequence* x. In modern terms, we thus have:

$$(d^2x)_n=(dx)_n-(dx)_{n-1} \text{ for } n\geq2 \text{ and } (d^2x)_1=dx_1\,(=1)$$

2   'Philosophie et mathématiques leibniziennes', in [Granger 1994], 223.

Hence:

$$(dx_n) = \left(\int (d^2 x) \right)_n = 1 + \sum_{2 \leq k \leq n} (d^2 x)_k = 1 + \sum_{2 \leq k \leq n} \left[(dx)_k - (dx)_{k-1} \right]$$

$$dx_n = 1 + dx_2 - dx_1 + \sum_{3 \leq k \leq n} -\frac{1}{k(k-1)} + \frac{1}{(k-1)(k-2)}$$

$$= -\frac{1}{2} + \sum_{3 \leq k \leq n} \frac{2}{k(k-1)(k-2)}$$

Since the sequence dx_n also converges to zero, one thus has: $\displaystyle\sum_{k \geq 3} \frac{2}{k(k-1)(k-2)} = \frac{1}{2}$.

As to the sum of the reciprocals of the pyramidal numbers, one then gets:

$$\sum_{k \geq 3} \frac{6}{k(k-1)(k-2)} = \frac{3}{2}$$

This produces the result provided by the first method (see above, Chapter VIII-D).

Descartes and Leibniz, numbers *versus* functions

In his second construction of the triangles, Leibniz had therefore considered the lines of a triangle and not its elements. Furthermore, this approach had proven to be far more successful than the first one. To deal with lines, however, was to manipulate sequences of numbers (in the present case, either integers or reciprocal integers). However, sequences are just a special case of functions, one in which the definition set is that of the integers (IN).

The (epistemologically new) point of view introduced by Leibniz to this issue was thus *functional* in essence. In this respect, he again opposed Descartes' designs. Leibniz indeed constantly put forward the distinction between 'the algebra' of Descartes and his own (infinitesimal) 'calculus' (see, for example, GM V 350). Yet an important aspect of this distinction lies, for us today, in the fact that Descartes considered that the objects of his algebra were numbers – it was thus a numerical calculation. On the other hand, however, since Leibnizian differentiation and summation necessarily apply to functions, the whole Leibnizian calculus must be considered as a functional calculation (concerning functions) or a sequential calculation (concerning sequences or series, which are special cases of functions). Thus, for example, in a product with the symbolization x.y, the letters stand for a product of numbers in Descartes' algebra, while they very often stand for a product of functions in the calculation of Leibniz, for example $d(x.y)$, or $\int x, y$. Moreover, this is what is at work in his second construction of the triangles. As we know, this distinction between the numerical conception and the functional design later became a crucial epistemological component, coextensive with the subsequent development of mathematics – it could not, however, be brought to light as such by the protagonists of the time.

IX-B THE 'SUM OF ALL THE DIFFERENCES',
FROM LEIBNIZ TO LAMBERT AND LEBESGUE

The 'sum of all the differences' in a contemporary environment

In the previous section, I proposed a reconstitution of the thought of Leibniz common to the two triangles, based on the method of "sum of all the differences" (hereafter denoted "S * d"). I shall expose, now, a modern direct version of the same result; it is extended within the context of normed vector spaces. Such a property that connects any numerical sequence to a certain associated series is commonly used today:

Theorem: Let E be a normed vector space, and u_n $(n \geq 1)$ the general term of a sequence of elements of E. We associate a series of elements of E by its general term $z_n = u_n - u_{n-1}$ $(n \geq 2)$ and $z_1 = u_1$. *As such, the nature of the u_n-sequence (i. e., its convergence or its divergence) is the same as that of the z_n-series.*

Moreover, in the convergence case, if $\lim u_n = \lambda$, then $\sum\limits_{n \geq 1} z_n = \lambda$

Proof:
The proof is immediate by faithfully following Leibniz's procedure, that is:

$$z_1 = u_1; \ z_2 = u_2 - u_1; \ z_3 = u_3 - u_2; \ \text{etc.,} \ z_{n-1} = u_{n-1} - u_{n-2}; \ z_n = u_n - u_{n-1}$$

Thus $\sum\limits_{k=1}^{k=n} z_k = u_n$. Accordingly, the series converges if, and only if, the sequence converges and the limit of the sequence is the sum of the series.

Note that the same property remains valid if one starts from any integer p, $p \geq 2$.

Lambert, continued fractions, and the 'sum of all the differences'

In the eighteenth century, Euler[3] and Lambert made extensive use of the 'sum of all differences' in the context of continued fractions[4].

I limit myself, here, to the approach of Lambert, which is simultaneously complicated and interesting. In his famous paper on the irrationality of π ([Lambert 1761]), he wanted to develop tan V in a generalized continued fraction (V is a rational number)[5]. For this purpose, Lambert operated on a certain infinite continued fraction (let us call it G), of which he would naturally first check that it is convergent. He thus examines the sequence of the convergents of (G). These successive convergents are a sequence of rational fractions with respect to W (W = 1/V), which he denotes F_n (W). To establish the convergence of the *sequence* F_n, he sub-

---

3   For instance, in [Euler 1835] Vol. I, § 363–368. See also [Euler 1744], § 8 and 9.
4   To my knowledge, however, neither Lambert nor Euler made reference to Leibniz on this point.
5   At the end of his report, Lambert will propose an important definition of the transcendent numbers (see infra Chap. XIII-B).

sequently used a geometric majoration[6] in order to prove that the *series* of the general term $F_n - F_{n-1} = z_n$ is convergent. Next, he used "S * d", to conclude as to the sequence F_n: this is exactly the method of Leibniz. Here, Lambert (§ 32, p. 130):

> It is a question of knowing the law according to which the fractions [...] approach the value of the tangent. To this end, we'll just subtract each one of the one that follows (...)

(S*d) is now a common method possible in both directions, namely from the sequences towards the series, or vice versa. Using the method of Leibniz, Lebesgue[7], for example, studied the convergence of an *arbitrary* unlimited continued fraction (L):

$$\cfrac{a_1}{b_1 + \cfrac{a_2}{b_2 + \cfrac{a_3}{b_3 + \ldots}}}$$

He denotes as $F_n = \dfrac{P_n}{Q_n}$ the convergent of order n of (L), and shows by "S*d" that the sequence converges if, and only if, the numerical series with general term:

$$F_n - F_{n-1} = \frac{(-1)^n a_1 a_2 \ldots a_n}{Q_{n-1} Q_n}$$

is convergent. He is then able to give important general criteria on *the convergence of any unlimited continued fraction.*

Note, also, another important application of the 'sum of all the differences' to the convergence of some numerical series; it is now known as 'Abel's rule and transformation'[8]. This subtle rule applies to enable us to decide as for the convergence of some numerical series that are not absolutely convergent, as for some trigonometric series[9].

---

6   See our study of Lambert's report in [Serfati 1992], 62–83.
7   [Lebesgue 1950], 98.
8   This remark was suggested to me by B. Gagneux. For a modern presentation of Abel's rule, see [Serfati 1995a], p. 13 for the statement, and p. 44 for the proof of the result.
9   See [Serfati 1995a], 43–44.

IX-C ON THE HARMONY OF THE HARMONIC TRIANGLE

Leibniz and Mengoli

Now we return to some interesting historical aspects. Contrary to what he could have believed at the time, Leibniz was not the first to have introduced the harmonic triangle. I will briefly develop, here, a priority point regarding Leibniz and Mengoli.

The research areas of Leibniz and Pietro Mengoli (1625–1686) crossed repeatedly, particularly on two central issues at stake in this chapter: first the sum of the reciprocals of the figurate numbers, and second the harmonic triangle. An article by Siegmund Probst[10] concerning the reception by Leibniz of the work of Mengoli provides valuable historical information – in this section, I will widely rely on its conclusions.

In the last months of 1672, Leibniz had found the sum of the reciprocals of the figurate numbers by using his first method (see above, Chapter VIII-D). At that time, he did not know Mengoli's work. Shortly afterwards (January/February 1673), he was on a mission in London and he learned, from Pell and Oldenburg, that Mengoli had preceded him on this subject, in a book published in 1658 in Bologna, the *De additione fractionum sive quadraturae arithmeticae*. Later on (1672), Mengoli published another book, the *Circolo*, where he introduced for the first time in history (and therefore before Leibniz) the harmonic triangle (*Circolo*, p. 4 – see figure below), which he called 'terza tavola triangolare'. The objective of Mengoli, however, was not the study of the properties of the triangle *per se*, as Leibniz would do, but to use it to evaluate – by means of approximations – some quadratures of curves and, particularly, the squaring of the circle, through what would be written

today as $I = \int_0^1 \sqrt{x(1-x)}\,dx$ [11].

---

10   [Probst 2007].
11   In fact, if one has $y = \sqrt{x(1-x)}$, then $x^2 + y^2 - x = 0$, and $0 \le x \le 1$. The issue is therefore the area of the portion of the plan limited by a semicircle (that is, half a disk). In [Massa Esteve 2009] – an interesting article devoted to the mathematical work of Mengoli – the authors point out that the real ultimate goal of Mengoli was to construct a 'harmonic triangle of interpolation'. Such a triangle was even more general than the harmonic triangle, and its standard element, expressed in modern terms, was represented by a certain numerical value of the Euler Beta function: $\beta(p,q) = \int_0^1 t^{p-1}(1-t)^{q-1}\,dt$ This integral is convergent if p>0 and q>0, and is used here for some half-integer fractional values of p and q (page 336). For instance, the integral:

$$\beta(\tfrac{3}{2},\tfrac{3}{2}) = \int_0^1 t^{\frac{1}{2}}(1-t)^{\frac{1}{2}}\,dt$$

provides the quadrature of the circle.

9   E denominando l'vnità, per ciaícuno di quetti termini,
facciaſi la terza tauola triangolare delle parti, che ſegue.

$$
\begin{array}{ccccccc}
& & & 1 & & & \\
& & 1(2) & & 1(2) & & \\
& 1(3) & & 1(6) & & 1(3) & \\
1(4) & & 1(12) & & 1(12) & & 1(4) \\
\end{array}
$$

1(5)   1(20)   1(30)   1(20)   1(5)

1(6)   1(30)   1(60)   1(60)   1(30)   1(6)

1(7)   1(42)   1(105)   1(140)   1(105)   1(42)   1(7)

1(8)   1(56)   1(168)   1(280)   1(280)   1(168)   1(56)   1(8)

'Terza tavola triangulare' in the Circolo of Pietro Mengoli (1672)[12].

Leibniz and the harmony of the triangle

Thus, Leibniz did not know Mengoli's 'terza tavola' when he created his own harmonic triangle between late 1673 and mid–1674[13]. Moreover, it was thus Leibniz himself, and he alone, who named this triangle 'harmonic'[14]. When we measure the centrality of the concept of harmony in Leibniz's metaphysics, his introduction of the term in a mathematical context is obviously fully meaningful. We return later to this question.

The major conclusions of the previous Section IX-A can thus be expressed as follows: the same pair of operations, the signs of which are d and $\int$, is at work on sequences and series of any type whatsoever. Furthermore, these operations are mutually reciprocal. This organization was essential because, for Leibniz, the strength of such a situation of reciprocity was based on the fact that it is originally coextensive with the *dissociation of identity* (see *supra* IX-A).

The present conclusions emphasize the importance of a reciprocity in act in every triangle – harmonic or arithmetical. However, and on the other hand, what would have been completely 'harmonious' for Leibniz in the harmonic triangle rested on its comparison with the arithmetical triangle: by its simple presence and its properties, the harmonic triangle *demonstrated* that a mathematical situation which we could consider to be restricted to some properties of a combinatorial type concerning, in the arithmetical triangle, simple sequences of integers, was valid just

---

12   Note the particular notation of Mengoli for fractions. He denotes, for example, 1(12) for $\frac{1}{12}$.

13   The oldest text of Leibniz on the triangle is probably 'De Triangulo harmonico', AVII, 3, N. 30, p. 336 (dated late–1673 to mid–1674). Probst explains (p. 7) that Leibniz had knowledge of the work of Mengoli only later on, in April 1676.

14   Leibniz claimed for himself the invention of the term 'harmonic triangle' in these terms: "Porro Triangulum hoc voco Harmonicum quemadmodum clarissimus quondam Geometria Blasius Pascalis suum vocabat Arithmeticum (…)" (see [Parmentier 2004], 234). See also 'Origo inventionis trianguli harmonici' (AVII, 3 N53₃, 712).

as much for sequences of a radically different type – in fact, the opposite – first made with reciprocal integers, then decreasing, and with zero as the limit[15].

For Leibniz, the permanence of one and a single operational pattern, valid in two contexts, so opposed, established the validity in itself of the pattern, with even more force. We recognize, here, one of the main laws of the Leibnizian harmony which we recalled in Chapter VIII-A, namely: the greater the diversity, and the unity in diversity, the greater the harmony. Let us repeat: for Leibniz, two reasons were conjugated to assert the validation of the pattern as a universal tool in the calculation of sequences and series. At first, the complete opposition in essence of the two triangles. Next, the equal success of an identical pattern $\left(d - \int \right)$ when applied to both triangles and their associated series.

This epistemological pattern even appeared, for a time, so general and operative in the eyes of Leibniz that he went as far as believing that he had found in this model a way of summing all numerical series[16]! In fact, the method obviously allows us to sum series which are themselves already differences. As such, it was a mistake on the part of Leibniz; we may, however, consider it understandable, given the epistemological importance of the scheme he had cleared. Later still, through his principle of continuity, the pattern became, for him, a higher principle of universal calculation (see the following Section IX-D).

IX-D FROM RECIPROCITY TO THE "NEW CALCULATION OF THE TRANSCENDENTS". THE "CONSIDERATIONS (...)" OF 1694. THE *PATTERN*.

A new calculation with eight pairwise reciprocal operations

The discontinuous field – that of sequences and series – was thus provided with an additional pair of reciprocal operations, with the signs $\int$ and d. Equipped with his – metaphysical – principle of continuity[17], Leibniz could certainly hope that the same pair would apply equally well to geometry and to the continuous field – that of functions. The consideration of such an extension of reciprocity gradually led him to upset the structure of the former calculation (that of Vieta and Descartes), as well as to design, in several stages, a new pattern. This elaboration was developed, at the symbolic level, according to various phases that I have described in detail in other texts ([Serfati 2008a] and [Serfati 2010a]). I shall here limit myself to summarizing the main clauses.

---

15 Let us underline a similarity in the field of continuous quantities: one can conceive an extension of the summation procedure from the case of polynomials to the one of inverses of polynomials (i. e., rational fractions).

16 S. Probst: "Mit dieser Methode, die bei der Summierung der Folgen der reziproken figurierten Zahlen zum Erfolg führte, glaubt Leibniz, als erster eine universelle Methode zur Reihensummierung gefunden zu haben," [Probst 2006], 172.

17 See our detailed commentary on such a mechanism and on its symbolic foundation in [Serfati 2010a]. See also [Philonenko 1967].

Having established that differentiation and summation were reciprocal operations[18], Leibniz allowed himself to consider d and $\int$ as a pair of fully-fledged assemblers, just like were sum and difference on one side, product and quotient on the
other hand, and also powers and roots. To the six operations of Descartes' algebra,
Leibniz therefore considered that he himself had added a new pair of two reciprocal
operations; he thus constituted a frame of calculation that was his own, with eight
operations, encompassing and surpassing the former scheme.

On the other hand, d and $\int$ possessed very different fields of existence – as
Leibniz himself points out (cf. GM V 308). The new calculation included this very
particular feature: differentiations could be performed almost everywhere, even
with irrationals (Leibniz insists: irrational numbers will never be a source of embarrassment when it will be a question of differentiating)[19]. On the other hand, actual
summations are – very usually – impossible[20].

Given the symbolic conceptions of Leibniz, being able to add to the calculation a
new assembler $\int$ certainly represented, for him, a fascinating process. Yet what could
he make of such an operator if he could not use it to freely calculate? Such inaptitude
was certainly not unacceptable *a priori*, as Leibniz himself observes (see below): it
already existed, indeed, in the pair powers-roots of the former calculation without
compromising the strength of the construction. It is always possible to raise an integer
or a rational number to any power. However, one cannot generally extract the root (for
example, a simple square root) of such a number, by remaining within the field of the
rational numbers (with the obvious – and 'rare' – exception of an integer which is itself a square). The introduction of the irrational numbers then came to fill this
gap. Similarly, it is always possible to perform summations in the exceptional case of
polynomials, that is to say, to make explicit the results with the previous symbolism;
the value is still a polynomial. However, Leibniz was soon confronted with the absence of a similar explication in a very significant number of other summations. On
this point, the case of rational functions was certainly very instructive for him.

18 Leibniz's interpretation of the symbolism of reciprocity sometimes proves very modern. Thus
    he writes to the Marquis de l'Hôpital: "For example, I found, as $x^{-1} = 1 : x$, as well $d^{-1}x = \int x \gg$,
    (GM II, 274).
19 See, for example, the detailed title of the founding article of Leibniz in *Acta Eruditorum* of
    1684: '*Nova Methodus pro Maximis et Minimis, itemque Tangentibus, quae nec fractas, nec
    irrationales quantitates moratur et singulare pro illis calculi genus*' (GM VII, 326), i.e., a
    '*New method for maxima and minima, and for tangents, that is not hindered by fractional or
    irrational quantities, and a singular kind of calculus for the above mentioned.*'
20 This point is highlighted by Breger: "This is not surprising; tangents to algebraic curves are
    easily algebraically calculated, but it is not often the case for the quadratures of algebraic
    curves. The difficulties that arise because of this fact for the formulation of the integral calculus
    were the main motivation for Leibniz to be interested in a comprehensive manner to the prob-
    lems of the transcendent," [Breger 1986], 127.

From Descartes to Leibniz, the rational functions

The rational functions are very common and simple, of one degree only more complex than the polynomials. However, except in a small number of exceptional cases, the sum of a rational fraction is *not* a rational fraction. In other words, the summation of an expression of this type, written in the symbolic system which served to constitute it, cannot in general be carried out in this system. In what follows, I at first focus on the foundations of this affirmation.

Leibniz and the partial fraction expansion of the rational fractions

This result is demonstrated through a general method of integration of any rational function $\dfrac{P(x)}{Q(x)}$ with real coefficients; this method, now standard, is called *partial fraction decomposition* (or *expansion*). In such a decomposition – over the field of real numbers – there are, putting aside the case (if any) of a polynomial, only two kinds of partial fractions, the form of which depends on the factorization of the denominator. The fractions of the first kind, that is $\dfrac{M}{(a+x)^k}$, give, by integration, rational functions (if $k \geq 2$) and logarithms (if $k = 1$). In Leibniz's terminology, the procedure that involved logarithms fell within the quadrature of the hyperbola[21].

The fractions of the second kind, namely $\dfrac{px+q}{\left(ax^2+bx+c\right)^k} = \dfrac{P(x)}{\left(Q_1(x)\right)^k}$, where Q_1 is without real roots, give, by integration, functions involving logarithms and the function Arctangent. In Leibniz's terminology, the procedure fell – this time – within the quadrature of the circle[22]. Accordingly, there is no primitive of a rational fraction that is still a rational fraction – except a few exceptional cases. In general, such a primitive involves simple or compound forms of logarithms and/or Arc tan,

for example, as $\ln\left(\dfrac{1+x^2}{1-x^2}\right) = A(x)$ or else Arc tan $(2x+1) = B(x)$[23]. From this, it also

follows that two transcendent quadratures, and only two (those of the circle and the hyperbola), are sufficient to explicit the quadrature of any rational fraction.

Given the frequency of use of rational fractions, this result is essential, and one understands that Leibniz was always interested in the quest for such a general method. He devoted to this subject an interesting paper in the *Acta Eruditorum* of 1702, namely the *Specimen Novum Analyseos Pro Scientia Infiniti Circa Summas et*

---

21   GM V, 358–359.
22   *Idem.*
23   A and B are obtained by summation of rational functions. We indeed have:
$$A(x) = \int \frac{4x}{1-x^4}dx \quad \text{and} \quad B(x) = \int \frac{dx}{2x^2+2x+1}$$

Quadraturas[24]. In this text, however, he makes a mistake due to an inadequate analysis of the structure of the factorization of the denominator $Q(x)$ in a number of specific cases (e.g., when $Q(x)$ is a multiple of (x^4+a^4)). He therefore concluded (incorrectly) as to the insufficiency of the aforesaid two basic quadratures[25].

Leibniz and the primitives of the rational fractions

Thus, while the rational fractions had been typically 'Cartesian' as to their symbolic formation, their summation would undoubtedly exceed Descartes, to become a genuine 'Leibnizian' problem. We refer, here, to what we outlined above in II-A. What about, for example,

$$\int \frac{1}{1+x} \quad \text{or} \quad \int \frac{1}{1+x^2} \quad \text{or} \quad \int \frac{1-x}{x^3+x} \ ?$$

Leibniz was well aware that the first two examples correspond in geometry to two crucial problems of quadrature – of the hyperbola and of the circle, respectively – well known since Antiquity, widely studied, and considered impossible until the mid-seventeenth century. As mentioned above, in II-A, these two problems, however, were later resolved: in 1668 by Mercator in *his Logarithmotechnia for the hyperbola*[26], and then by Leibniz himself in 1673 for the circle. These summations presented (regarding their statement) an immediate geometrical interpretation, namely the calculation of a certain plane area. However, regarding their values, Leibniz – contrary to Descartes – was able to calculate them; they were, in fact, given in the form of the sum of some infinite numerical series obtained through the term-by-term integration of certain power series (see above, II-A). The methods for quadratures, and therefore those for summations, fell within some infinitary frame, that of a new analysis. Leibniz then departed from this situation to convert it into an unprecedented conceptualization: the summation of an arbitrary rational function, far from being meaningless, gave birth to a new type of object: *a transcendent expression*[27].

---

24   GM V, 350–361.
25   The factorization of x^4+a^4 in *real* polynomials of the second degree is
$$x^4 + a^4 = \left(x^2 + a^2\right)^2 - 2x^2a^2 = \left(x^2 + a\sqrt{2}x + a^2\right)\left(x^2 - a\sqrt{2}x + a\right), \text{ and we are returned to}$$
partial fractions of the second kind, an already envisaged case. Leibniz, wrapped in imaginary quantities, does not reach this result. In fact, the quadratures of the circle and of the hyperbola are thus enough. See on this point the interesting comment of [Parmentier 2004], 383.
26   [Mercator 1668].
27   The end of this section is widely based on the conclusions of [Breger 1986].

'Ordinary' Analysis
versus
'New' Calculation of the Transcendents

These ideas of Leibniz found their highest expression in a 1694 article, *Considérations sur la Différence qu'il y a entre l'Analyse Ordinaire et le Nouveau Calcul des
Transcendantes*. This text of his maturity (Leibniz was forty-eight years old) is
significant for the completion of his conception of the transcendent – from then on,
at the same time symbolic and essential. Here is Leibniz:

> In the ordinary Analysis we can always free the calculation *a vinculo* and of roots by means of
> the powers, but the public has not yet the method to free it from the powers involved by means
> of pure roots. Similarly, in our Analysis of transcendents, we can always free the calculation *a
> vinculo* and of sums by means of the differences, but the public has not yet the method to free it
> from the differences involved by means of pure sums, or of quadratures: and as it is not always
> possible to actually extract the roots, in order to reach the rational magnitudes of the common
> Arithmetic, it is not always possible either to actually give the sums or the quadratures, in order
> to reach the ordinary or algebraic quantities of the common analysis. However, by means of the
> infinite series we can always express broken magnitudes as integers, incommensurable ones
> as rational, and transcendent as ordinary. And I gave a general way there according to which
> all the problems, not only those of differences, but still of differentio-differentials, or sums of
> sums, and beyond, can be sufficiently built for the practice: as I have also given a general con-
> struction of the quadratures by a continuous and adjusted movement.
>   Finally, our method being strictly this part of the General Mathematics that deals with
> the infinity, it is what makes that we strongly need it when applying Mathematics to Physics,
> because the character of the infinite Author usually enters the operations of the nature (GM V,
> 308).

A truly general calculation

The previous lines clearly explain what was – ultimately – the objective sought by
Leibniz: to found a universal calculation (i. e., generalized, compared with those of
Descartes and Vieta) and the means of reaching this goal.

The final stage for Leibniz took the form of a new extension of the field of the
transcendence (beyond even the *Considerations*, which implicitly dealt only with
rational fractions): it was a universal extension of the summation procedure, from
the rational fractions up to arbitrary functions. Thus, Leibniz could consider it legitimate that no summation was impossible, provided one would accept that the
calculation was no longer a matter of the former "Ordinary Analysis", but from now
on of a new mathematical field, strictly Leibnizian, the "New Calculation of the
Transcendents", of which Leibniz himself, at the same time the inventor of the
sign $\int$ and the master of the symbolic substitutability, became *ipso facto* the full
creator; he thus exceeded Descartes[28] and Newton[29].

---

28 GM V, 124.
29 See Leibniz's various comments on the subject of mathematics in Descartes, and also in New-
    ton in [Serfati 2005], 336–337.

Transcendent expressions are exactly those that are impossible to explicate

In fact, in order to base his argumentation on the legitimacy of his new objects, Leibniz put forward a dialectic of the impossible, which is both interesting and analogical, while simultaneously paradoxical. Here is Leibniz in another famous 1710 text, *Symbolismus Memorabilis* (...)[30]:

> In the same way as the impossibility to extract roots in the rational numbers leads to irrational magnitudes, in the same way the impossibility of summing algebraic expressions leads to transcendent magnitudes.

Breger comments, very rightly, on[31]:

> The legitimacy of such objects results only from the fact that they went out of a problem, which is not soluble in an algebraic way.

Thus, for Leibniz, these transcendent objects derive their existence from the impossibility of their clarification – that is to say, of their absence in the Cartesian field! This statement appears to us as an attempt at a definitive dissolution of the 'riddle of the non-Cartesian field' – finally brought to light by Leibniz. As we have seen repeatedly in this volume, the latter had vainly tried to inventory (in a comprehensive and reasoned way) the objects of non-Cartesian mathematics. The problem will obviously have been – as much for Leibniz as for his successors – that such a definition, entirely negative, does not allow us to apprehend these objects by themselves. Their designation by a convenient – but general – term as 'transcendent' had certainly constituted an interesting conceptual stage, but it did not advance their positive analysis. The paradoxical definition above allows us, on the contrary, to believe in a slightly more concrete symbolic approach: for Leibniz, from then on, those expressions obtained by summation from explicit Cartesian expressions were transcendent, on the express condition that the outcome itself of this summation was not explicable in the Cartesian mode.

The shift from the *Ordinary Analysis* towards the *New Calculation of the Transcendents* thus embodies – definitively, in my opinion – the major historical and epistemological evolution of Leibniz's mathematical thought, at the same time as his completed triumph over the conceptions of Descartes. One understands to what extent his approach was incompatible with Descartes' mathematical thought!

---

30   GM V, 377 = 'Symbolismus Memorabilis Calculi Algebraici et Infinitesimalis in Comparatione Potentiarum et Differentiarum, et de Lege Homogeneorum Transcendentali'. *Miscellanea Berolinensa*. 1710.

31   [Breger 1986], 127.

The advent of the transcendence and the detachment
from geometry as the only guarantee of the truth

From then on there emerged, in Leibniz and then in the mathematical community, symbolic forms devoid of any immediate geometrical sense; they were mainly obtained by the pure combinatorial play of a succession of substitutions, such as:

$$\int 6\left(\frac{x^3+2x^2-1}{4x+1}\right)$$

or

$$\overline{\int mx+n, \ dx\sqrt{hx^3+ixx+kx+l}}\ ^{32}$$

Alternatively, in a letter from John Bernoulli to Leibniz:

$$\int \overline{z^e d^m n} + e\int \overline{z^{e-1} d^{m-1} ndz}\ ^{33}\text{n}$$

Such symbolic forms were considered formally legitimate, and they were actually brought forward by Leibniz and the Bernoulli brothers; they discovered a cascade of new objects and problems – mechanically produced – without the need for a reference to a possible meaning. Let us repeat: the symbolic forms above had no immediate sense according to the concepts of the time, that is to say, *no immediate geometrical sense*[34].

It was an important moment in the history of the mathematical ideas following Descartes. From then on, indeed, geometry no longer worked as a privileged domain of the reality, which created the calculation, and controlled it at every step, but only as a distant (and vague) theoretical possibility of welcoming the return of the symbolism, at first rarely done *de facto*, and finally often abandoned *de jure*. Accordingly, the power of symbolism put the final touch to the historical detachment from geometry as the only guarantor of the truth. Thus, this episode in the history of mathematical ideas – the advent of the differential calculus and, correlatively, of the transcendence – corresponded at an epistemological level to a side-lining of the former geometrical world. This split was therefore coextensive with the development of symbolic writing. Whatever it might be, that each new symbolic structure could be opened to substitutions became a key demand, occurring in Leibniz's writings on every page. Leibniz thus rightly appeared proud of his 'Calculation of the Transcendents'.

Commenting on another essay of Leibniz, entitled 'Nouvelles remarques touchant l'analyse des transcendantes, différentes de celles de la Géométrie de M. Descartes'[35] (i.e., 'Further remarks concerning the analysis of the transcendent, distinct from those of Mr Descartes' Geometry'), Breger pertinently notes:

---

32 James Bernoulli to Leibniz, 28 February 1705, GM II, 97.
33 John Bernoulli to Leibniz, 20 (30) April 1695, GM II, 171.
34 In the words of Breger, "What was new in Leibniz's infinitesimal calculus was that calculations became somewhat independent of geometry", ([Breger 2016], 153).
35 GM V, 278–279.

However, the true mathematical content of this essay is not on the technical plan – it is much more a question, for Leibniz, of legitimizing transcendent expressions[36].

Harmony is restored: transcendence is no longer an obstacle to the calculation but an opening towards a new mathematics

To conclude this section: Leibniz did not limit himself to observing the impossibility of certain summations. Accepting this impossibility would have meant that these concatenations of signs represented nothing – or nothing more than themselves – or else that there was no other interpretation for them than the fact that they were summations. On the contrary, Leibniz claimed this statement – strictly inconceivable for Descartes – that these summations gave birth to expressions obtained by pure symbolism, and (usually) of an entirely new type, which he called *transcendent*[37].

From then on, it became possible to sum any expression and harmony was thus organically (and fully) restored, that is to say, $\int$ was no longer an obstacle to the calculation, but rather an opening towards a new mathematics.

For Leibniz, however, this transcendent analysis always remained correlated to its validity in geometry, in which the reciprocity between both operations is embodied in what is called the Fundamental Theorem of Analysis[38]– that is to say, the reciprocity between the problem of tangents and the inverse problem of tangents, the latter being directly linked to that of quadratures. This correlation of validity between symbolism and geometry was the subject of many reviews (cf. for example [Breger 1986]). For Leibniz, its justification – metaphysical – would be coextensive with the use of the principle of continuity (see [Serfati 2010a] and [Duchesneau 2008]).

The Leibnizian *scheme*: a calculation with reciprocity and iterations

From the diversity of situations, namely, the arithmetical triangle, the harmonic triangle, arbitrary real sequences, equations of algebraic curves, and finally equations of transcendent curves, thus resulted the permanence of a calculation scheme with reciprocity and iterations that in other texts I simply call 'The Scheme' (see [Serfati 2011a]).

---

36   [Breger 1986],128.
37   With modern definitions (see details, infra, chapter XV), a function F (of one variable) is algebraic if, and only if, it is rational. A non-algebraic function is called 'transcendent'. About the various conceptions of transcendence for functions in Leibniz, cf. the detailed study of [Breger 1986]. He observes (p. 130): "To be complete, it should be noted that Leibniz uses the concept of 'transcendent' in a sense that is not quite that of today." The classification of Euler [1748] accepts the radicals in the definition of algebraic functions, but distinguishes between rational algebraic functions and irrational algebraic functions. We detail this below in Chapter XIII. See also our study [Serfati 1992], 49–50.
38   On this important point, see for example [Wahl 2011], 1180.

IX-E MATHEMATICAL COMPLEMENTS.
ON THE PROPERTIES OF THE HARMONIC TRIANGLE

The harmonic triangle was a remarkable mathematical creation of Leibniz, to which it seems to me interesting to devote an additional mathematical study. I develop here some aspects that were not examined in the seventeenth century. I shall base my study on two 'modern' sources, namely ([Henry 1881] and [Serfati 1967].

Theorem 1[39]: For $n \geq p \geq 1$, we define $T(n,p) = \sum_{p \leq j \leq n} K(j,p)$.

Then, for every $p \geq 1$, we have $\lim_{n \to \infty} T(n,p) = \dfrac{1}{p}$.

Remark: T(n, p) is the sum of the elements in the same column (number p), from the diagonal element, and down to any element of the harmonic triangle (see figure below).

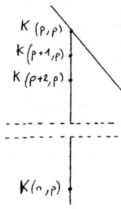

The limit of T (n, p), if it exists, is thus the sum of the convergent series of all the elements of the same column. Recall that, for the *similar* sum of the arithmetical triangle, by putting $U(n,p) = \sum_{p \leq k \leq n} C(k,p)$, we have

U(n,p) = C(n+1, p+1).

Proof: Let K(n, p) be the element of the n-th line and the p-th column of the harmonic triangle (cf. supra VIII-C). We have:

K(p, p) = K(p–1, p–1) – K(p,p–1)
K(p+1, p)= K(p, p–1) – K(p+1, p–1)
K(p+2, p) = K(p+1, p–1) – K(p+2, p–1)
etc.

---

39   [Serfati 1967],126. This statement is that of a problem which I had made up for my students and then published in a volume in collaboration.

$K(n–1, p) = K(n–2, p–1) – K(n–1, p–1)$
$K(n, p) = K(n–1, p–1) – K(n, p–1)$

The method is still the 'sum of all the differences'. By adding, there are only two terms:

$$T(n,p) = K(p–1, p–1) – K(n,p–1) = K(p–1,0) – K(n, p–1)$$

$$= \frac{1}{p} - \frac{(p-1)!}{(n-p+2)...n(n+1)}. \text{ Hence } \lim_{n \to \infty} T(n,p) = \frac{1}{p}$$

Theorem 2[40]: For $n \geq 1$, we define $S_n = \sum_{0 \leq j \leq n} K(n,j)$.

Then, $S_n = \frac{S_{n-1}}{2} + \frac{1}{n+1}$ and the sequence S_n is convergent; its limit is zero.

Note: S_n is the sum of all the elements of a same line of the harmonic triangle. Recall that, for the arithmetical triangle, the corresponding sum is 2^n.

Proof:

$K(n, 0) = K(n, 0)$
$K(n, 1) = K(n–1, 0) – K(n, 0)$
$K(n, 2) = K(n–1, 1) – K(n, 1)$
$K(n, 3) = K(n–1, 2) – K(n, 2)$
etc.

$K(n, n–1) = K(n–1, n–2) – K(n, n–2)$
$K(n, n) = K(n–1, n–1) – K(n, n–1)$

By adding[41]:

$S_n = [K(n–1, 0) + K(n–1, 1)+...+ K(n–1, n–1)]–[K(n, 1)+...+K(n, n–1)]$

$S_n = S_{n-1} - [S_n - (K(n, 0)+ K(n, n))]$. Hence $2 S_n = S_{n-1} + (K(n, 0)+ K(n, n))$

$$= S_{n-1} + \frac{2}{n+1} \text{ (q.e.d.)}$$

Observe that $S_n > 0$, and then calculate:

$$S_{n-1} - S_n = S_{n-1} - \frac{1}{2}S_{n-1} - \frac{1}{n+1} = \frac{1}{2}S_{n-1} - \frac{1}{n+1}$$

If $n \geq 2$, then $n–1 \geq 1$, so that S_{n-1} contains at least both terms below:

$$= K(n-1,0) + K(n-1,n-1) = \frac{2}{n}.$$

---

40   [Serfati 1967], 126–128.
41   It is no longer, here, a 'sum of all the differences'.

Thus $S_{n-1} - S_n \geq \dfrac{1}{n} - \dfrac{1}{n+1} > 0$ and $S_n < S_{n-1}$.

The sequence S_n is decreasing and bounded below; it has therefore a limit $l\,(l\geq 0)$. To calculate l, we pass to the limit in the recurrence relation above: $l = \dfrac{1}{2}l + 0$. One then gets $l = 0$.

$$\text{Hence } \lim_{n\to\infty}\left[\sum_{0\leq j\leq n} K(n,j)\right] = 0$$

Theorem 3[42]: For $n\geq 1$, one defines $\Sigma_n = \sum_{0\leq j\leq n} (-1)^j K(n,j)$

$$\text{One has: } \Sigma_n = \frac{1+(-1)^n}{n+2}.$$

Remark: Σ_n is the alternating sum of the coefficients of a same line of the harmonic triangle. It should be recalled that the corresponding sum of the arithmetical triangle is zero.

Proof: First observe that if n is an ***odd*** integer (n = 2p+1), there is, in the sum, an even number (2p +2) of terms which are pairwise opposite:

$$(-1)^j\, K(2p+1, j) \text{ and } (-1)^{2p+1-j}\, K(2p+1, 2p+1-j)$$

Hence $\Sigma_{2p+1} = 0$.

If n is ***even*** (n = 2p), there are 2p+1 terms:

$$\Sigma_{2p} = K(2p,0)-K(2p,1)+K(2p,2) -\ldots+K(2p,2p) = U_p - U_p{'}$$

where $U_p = K(2p,0)+K(2p,2)-\ldots+K(2p, 2p)$

and $U_p{'}= K(2p,1)+K(2p,3)-\ldots+K(2p, 2p-1)$

A) To calculate U_p, we consider each of its terms as the first element of a difference:

$K(2p+1, 1) = K(2p, 0) - K(2p+1, 0)$
$K(2p+1, 3) = K(2p, 2) - K(2p+1, 2)$
etc.
$K(2p+1, 2p+1) = K(2p, 2p) - K(2p + 1, 2p)$

By adding, one gets:

$$A = U_p - B, \text{ and then } U_p = A + B =$$
$$K(2p+1, 0) + K(2p+1, 1)+ K(2p+1, 2)+\ldots+ K(2p+1, 2p+1)$$

Hence $U_p = S_{2p+1}$.

---

42   [Serfati 1967], 128–129.

B) To calculate U_p', we do the same:

$K(2p+1, 2) = K(2p, 1) - K(2p+1, 1)$
$K(2p+1, 4) = K(2p, 3) - K(2p+1, 3)$
$K(2p+1, 6) = K(2p, 5) - K(2p+1, 5)$
etc.
$K(2p+1, 2p) = K(2p, 2p-1) - K(2p+1, 2p-1)$.

By adding, one finds

$$C = U_p' - D, \text{ and therefore } U_p' = C + D =$$
$$K(2p+1, 1) + K(2p+1, 2) + K(2p+1, 3) + \ldots + K(2p+1, 2p-1) + K(2p+1, 2p).$$

U_p' is thus equal to S_{2p+1} deprived of its two extreme terms:

$$U_p' = S_{2p+1} - K(2p+1, 0) - K(2p+1, 2p+1).$$

By symmetry, $U_p' = S_{2p+1} - \dfrac{2}{2p+2}$, so that:

$$\Sigma_n = \Sigma_{2p} = U_p - U_p' = \frac{2}{2p+2} = \frac{2}{n+2}$$

Hence $\Sigma_n = 0$ if n is odd, and $\Sigma_n = \dfrac{2}{n+2}$ if n is even.

By bringing together the results, $\Sigma_n = \dfrac{1+(-1)^n}{n+2}$

Theorem 4[43]:

$$\text{For } n \geq 1 \sum_{0 \leq p \leq n} \frac{1}{K(n,p)} = (n+1) \cdot 2^n$$

Remark: This is the sum of the reciprocals of the terms of a same line of the harmonic triangle.

Proof: One has:

$$\sum_{0 \leq p \leq n} \frac{1}{K(n,p)} = \sum_{0 \leq p \leq n} (n+1)\binom{n}{p} = (n+1) \sum_{0 \leq p \leq n} \binom{n}{p} = (n+1) \cdot 2^n.$$

Examples: If n = 3, there are four terms and $(3+1)2^3 = 32 = 4+12+12+4$.

If n = 4, there are five terms and $(4+1)2^4 = 80 = 5 + 20 + 30 + 20 + 5$.

43 [Henry 1881], 6–7.

Theorem 5[44]:

$$\sum_{n \geq 2}\left(\sum_{k \geq 2} \frac{1}{n^k}\right) = 1$$

Remark: This is a property of the Riemannian numerical series, derived from the sum of the reciprocals of triangular numbers.

Proof: The basic relation is $a_n = \sum_{k \geq 2} \frac{1}{n^k}$, that is, the sum of a geometrical series with the ratio $\frac{1}{n}$, deprived of its first term $\frac{1}{n}$. We thus have $a_n = \frac{1}{n^2} \frac{1}{1 - \frac{1}{n}} = \frac{1}{n(n-1)}$.

It thus remains to calculate $\sum_{n \geq 2} a_n = \sum_{n \geq 2} \frac{1}{n(n-1)}$. This latter sum (the half of the reciprocal triangular numbers) has already been calculated by the method "S*d" applied to the harmonic triangle (see above, IX-A). The result is equal to 1.

---

44   [Henry 1881]. This is Corollary 1 (p. 8) of Henry; the announced result is true but the solution seems to contain diverse errors.

CHAPTER X
TRANSCENDENCE AND IMMANENCE.
SOME TERMINOLOGICAL MARKS
BEFORE AND AFTER LEIBNIZ

X-A ON NICHOLAS OF CUSA AND ON THE ORIGIN OF THE TERM
'TRANSCENDENT' IN LEIBNIZ

As we have said, the term 'transcendent' as introduced by Leibniz into mathematics had no theological or philosophical connotation for him. Therefore, it is obviously of interest to investigate its origin. It seems to me likely that he discovered it in the works of Nicholas of Cusa (1401–1464), who was a Cardinal and a mathematician[1].

Leibniz was familiar with the mathematical work of Cusa. As early as the Parisian years (July-August 1676), he mentioned it in the *De Studio Intentiore Geometriae*[2]. A year later, in 1677, he planned the writing of a note, entitled *Nicolaus Cusanus Egregie*, which, however, was not drafted. Leibniz wrote there only these few words: "*Nicolaus Cusanus Egregie: id est creare primae Menti, quod numerare est nostra.*" In 1682, in the course of an extended text on the history of science, the "*Constitutionem Scientiae Generalis*"[3], he again quotes Cusa about geometry. Still later (1696–1697), geometry and curves in Cusa, including the question of the priority of the invention of the cycloid, were the subject of an abundant exchange between Leibniz and Wallis[4].

On the other hand, the concept and the term transcendence are ubiquitous in Cusa's argumentation, especially in his major work *De docta ignorantia (On Learned Ignorance* – 1440)[5], where we usually find it involved in scientific metaphors, with the meaning of an absolute conceptual excess. Thus, in Book I:

> But *maximum* and *minimum*, as used in this [first] book, are transcendent terms of absolute signification, so that in their absolute simplicity they encompass – beyond all contraction to quantity of mass or quantity of power – all things[6].

Yet, and above all – and beyond these mathematical metaphors – this same concept of a supreme excess, registered in the term 'transcendence', is permanent in Cusa; it is, for him, fundamentally associated with the divine attributes. For Cusa, God

---

1 His name is often also spelled "Cues".
2 AVII, 6 N. 40, p. 434.
3 AVI, 4, N. 114, p. 487.
4 GM IV, 8–9, 12–13, 18, 20, 21, 27.
5 [De Cues 2010].
6 Book I, § 4 *The Absolute Maximum, with which the Minimum coincides, is understood incomprehensibly.* I used the English translation of Book I, made from *De docta ignorantia. Die belehrte Unwissenheit*, Book I (Hamburg: Felix Meiner, 1970, 2nd edition), text edited by Paul Wilpert, revised by Hans G. Senger.

must indeed be 'thought' in the terms of a perpetual excess. God is greater than all that we can conceive of him. Therein lies his transcendence. It is a conclusion on which an abundance of commentary agrees. For example:

> "Cusa considers that God is not only that we have to think of bigger, but also that it is what is bigger than anything that we can think of"[7].

The work of Cusa was very well known and frequently discussed in the seventeenth century. Descartes, for example, in a letter to Chanut of 6 June, 1647, evokes the ideas of Cusa as to the status of the infinite[8]. We therefore believe that Leibniz (he had, quite early on, knowledge of the work of Cusa in which mathematics and philosophy were closely imbricated) most likely borrowed this terminology of 'transcendence' to make it the specific mark – where no term was readily available – of an extreme 'excess' in mathematics.

X-B TRANSCENDENCE AND IMMANENCE.
MATHEMATICS AND PHILOSOPHY

The use of the word by Leibniz

If, thus, as we have assumed, Leibniz has borrowed the term 'transcendent' in Cusa to introduce it to mathematics, he did not load it with the philosophical/religious connotations and interpretations of which the Cardinal was an advocate. Initially, the word simply indicated what was, for Leibniz, outside the mathematical standards of the time, that is to say, 'outside the standard of the Cartesian geometry', or still, that which 'exceeded' (*transcended*) Descartes. However, as we have also seen, 'what exceeded Descartes' was then naturally extended to the symbolism, via mathematical expressions, and in a manner not directly connected to geometry: they were the letteralized exponentials, to which we saw that Leibniz was so attached, and by which he actually exceeded Descartes – as well as Newton – this time symbolically (see chap. III-C), not geometrically.

Leibniz seems to have had little use outside of mathematics for the term 'transcendent'. In particular, it is barely found in his philosophical work. As such, the Olms edition of the *Leibnizian Lexicon*[9], a dictionary which inventories the terms of Leibniz's vocabulary in accordance with the Gerhardt edition of the philosophical works of Leibniz, refers, using the word 'transcendent' (p. 355), to only three strictly *mathematical* occurrences of the term. Leibniz coined the term and so often used it in mathematics – would he have reserved its exclusivity for these? However, one might wonder whether this term, mathematically used by Leibniz, has not received later on – tacitly, as the case may be, or unconsciously – a philosophical

---

7   Isabelle Raviolo's comment in *Actu Philosophia,* on the French translation of *De la docte igno-rance*, by Jean-Claude Lagarrigue = http://www.actu-philosophia.com/spip.php?article401

8   ATV, 51·

9   [Finster 1988].

connotation which is in agreement with (or at least in relation to) its mathematical Leibnizian definition. This issue is the subject of the following section.

Transcendence: some philosophical definitions

Accordingly, I briefly review the articles "transcendence" and "transcendent" in two contemporary French philosophical dictionaries, namely Foulquié ([Foulquié 1969], pages 733–735) and Lalande ([Lalande 1927], pages 1143–1145) in order to examine – if they exist – philosophical uses of the term 'transcendent' which are connected or compatible with the mathematical conceptions of Leibniz.

For the verb '*Transcend* ', Foulquié gives as its definition 'To exceed a certain level or certain medium after having passed through. To be situated beyond,' and for '*Transcendent*', in a sense that he specifies to be philosophical:

> Which is situated beyond or outside the area concerned, and not of the same nature than it. The opposite is *immanent*.
> 1. More usually with an idea of superiority (...) God is transcendent to the world; the spirit is transcending the matter.
> 2. Sometimes without an idea of superiority: which is outside the considered domain.

In order to illustrate this second meaning of the word, Foulquié then offers an interesting epistemological example taken from Duhem[10]. On his side, Lalande proposes, for '*Transcendence*', this illustrated definition:

> Literally: which rises above a given level or a given limit (...) In particular, which does not result from the natural play of a certain class of beings or of actions, but which requires the intervention of an external principle and upper to this one. It is in this sense that we oppose the 'immanent justice' resulting from the natural course of events, to a 'transcendent justice' or 'transcendent sanctions', which would be of a different order and higher.

On immanence

Not surprisingly, both authors also agree on the term 'immanent' as the philosophical opposite of 'transcendent' (this point is also, in this context, briefly quoted by Breger). Foulquié (pp. 346–347) offers:

> As opposed to transcendent: which is interior to the being or to the object of thought in question. Immanent justice, immanent sanction = internal to the guilty actions themselves, in that they cause the punishment naturally without intervention of an external authority (...).

Lalande, who proposes similar comments (pp. 470–471), states:

> Is immanent to a being or to a set of beings what is included in them, and does not result within them from an external action. The 'immanent justice', the 'immanent sanctions' are those that

---

10   Foulquié, p. 734: "For whom sticks to the processes of the experimental method, it is impossible to declare true this proposition: *All physical phenomena are explained mechanically*. It is also impossible to declare it false. This statement is *transcendent* to the physical method" [P. Duhem. *La théorie physique*. 2nd edition, 425].

result from the natural course of events, without the intervention of an agent that could be distinguished from them.

The immanence of the Cartesian algebra, as opposed to the transcendence of the Leibnizian infinitesimal Calculus

Both authors thus agree on a point which can be, in my opinion, connected to the two aspects of mathematical transcendence in Leibniz, namely the functional and the numerical. As we have just seen it with our two philosophers, a transcendent entity is the result of those that precede it and / or give rise to it, although by being of a different nature, irreducible to them – in a register of the difference thus, and without necessarily presenting the connotation of superiority that one can sometimes find in theology (cf. Foulquié above, 'transcendent', Definition 2). In brief, would something be fundamentally transcendent (philosophically speaking) *which has not the same nature than that from which it results*. However, this is exactly the case of a transcendent function obtained through the integration of Cartesian expressions, or of a transcendent quantity obtained by Leibniz by the sum of an infinite series of rational numbers. Thus, the summation (i. e., the primitivation) of rational fractions is *not* an immanent operation. This Leibnizian meaning of the 'transcendent', therefore, belongs also to the register of philosophy. Immanence, coextensive with the Cartesian algebraic calculation, is thus opposed to the transcendence of the Leibnizian infinitesimal calculus.

We shall note in passing an inaccuracy of Lalande (p. 1144) who defines – certainly correctly – in the modern sense 'a transcendent number', but mistakenly credits Leibniz with it; according to him, the latter would have given the definition in the *New Essays*. However, we saw above (see Chap. VII) that it was not the case in reality, and that nobody before Lambert (see, *infra*, XIII-B) had given an acceptable definition of it.

FOURTH PART
THE RECEPTION OF THE TRANSCENDENCE

INTRODUCTION TO THE FOURTH PART

Introduced by Leibniz, the terminology of transcendence was resumed, as we shall see, by many of his correspondents and also by some influential mathematicians of the time (with the notable exception of Newton). The extreme diversity of the reactions of the protagonists suggests some embarrassment towards the strength and originality of Leibniz's process. This perplexity accompanied some hesitation on the part of Leibniz himself as regards the exact positive extension of a negatively defined concept (what we called, above, 'the issue of the inventory').

In Chapter XI, we merely analyse the features of the initial reception of the transcendence – essentially in the late seventeenth century. However, according to its historical importance and its mathematical complexity, a later chapter will be devoted to the reception of the transcendent by Huygens (Chapter XII). Further, in Chapter XIII, we will describe three of the main aspects of this reception in the eighteenth century, by Lambert, Euler and the Encyclopaedia. Finally, in Chapter XIV, we will analyse a particular system of ideas about transcendence in the nineteenth century, a hundred and fifty years after Leibniz, in August Comte's work – he was, as one knows, a mathematician as well as a philosopher.

CHAPTER XI
THE RECEPTION OF THE TRANSCENDENCE
BY THE CONTEMPORARIES OF LEIBNIZ

The ordinary correspondents of Leibniz at this time, such as Tschirnhaus, L'Hospital, Huygens and Johann Bernoulli, spontaneously resumed the terminology of the transcendence, sometimes with inflections of sense or misunderstanding; for example, in Craig or Huygens. Tschirnhaus was obviously an addressee favoured by Leibniz for the presentation of new theories which 'exceeded' Descartes. As for L'Hospital, if he limited himself to the differential calculus (and not integral) applied to the transcendent curves, he always stayed as close as possible to the genuine conceptions of Leibniz. Of all these correspondents, Johann Bernoulli was certainly the most interested and the most acute for the new concept, to which he devoted a detailed analysis.

The texts of the authors evoked in this chapter (Newton, Craig, Tschirnhaus, Sturm, Huygens, L'Hospital, Jean Bernoulli) all come from the article quoted from Breger[1], and are given by him in a footnote. We propose here a brief review.

XI-A THE RECEPTION BY TSCHIRNHAUS (1678–1682)

"Transcendentes, ut vocas ..."

Tschirnhaus was initially a convinced Cartesian; he was also the first addressee for the arguments of Leibniz on transcendence. Following the letter of May, 1678, that Leibniz sent him (cf. GBM, 377, discussed above in IV-D1), Tschirnhaus soon resumed the term in several of his correspondences[2]; at first, almost immediately, in his very long answer dated 1678[3], to the aforementioned letter of Leibniz. Tschirnhaus points out the term ("transcendentes, ut vocas") (p. 391), showing that it is probably for him a first introduction.

A correspondence a little later (May 27th, 1682), from Tschirnhaus to Leibniz (GM IV, 487–490), is also interesting in this respect. At that time, Tschirnhaus had once again returned to Paris because he hoped to be elected a member of the Academy of Sciences there. He succeeded (he was an associate member), in particular by the recommendation and the friendly support of Leibniz.

---

1   [Breger 1986], 131–132, n. 77 to 88.
2   See our study in [Serfati 2011c], 279–281.
3   GBM, 382–398 = A II, 1, N. 20 = LBr 943 Bl. 39–43.

In this letter, Tschirnhaus asserts, clearly and with great confidence, that he can determine "with the same facility the tangents to *all* curves, both analytical and not analytical", "that is to say, transcendent":

> I would hope that Mr. Leibniz had shown me in an example how he determines the tangents to $x^y + y^x = a$ (...) I thought, at another time, to the expression with which I determine the tangents to all non-analytical curves, that is transcendent curves, with the same facility as to analytical curves (GM IV, 487).

This extract shows that Tschirnhaus endorsed the Leibnizian concept of 'transcendent' as opposed to 'analytical' (the latter term being understood as 'possibly subject to a Cartesian symbolism'). This conception is, of course, directly imported from the words of Leibniz in his aforementioned letter of 1678 (GBM, 376). Tschirnhaus has therefore well understood Leibniz and the example of a transcendent curve (of equation $x^y + y^x = a$) which he proposes, in reply, is quite relevant. This equation indeed follows a similar transcendent expression ($y^x + x^y = a^{xy}$), which Leibniz had proposed to him in his letter of 13 May 1681 (GM IV, 485). At that time (1681–1682), Tschirnhaus thus seems to identify at first all non-analytical curves with those provided by equations involving letteralized exponentials; this was fully in line with the first conceptions of Leibniz at that time (see above IV D4, *The exponential utopia*).

At the end of his letter (p. 490), Tschirnhaus then tries – unsuccessfully, as we imagine it – to supply the general method which he had announced for the determination of tangents to the general transcendent curves. He obviously could not do otherwise than to encounter the same difficulties of inventory which Leibniz had to face: how to positively identify the nature of an arbitrary transcendent curve?

In fact, Tschirnhaus exposes (very briefly!) his process only on an example (page 490); yet it is about a really very simple algebraic curve, namely the circle of equation $y^2 + x^2 - 2ax = 0$ (= $F(x,y)$)! Starting from the equation of the curve, he proposes, heuristically, and without any proof, the construction of a ratio Z (a 'fraction', he says) such as that, for him, $Z + x = \sigma$, where σ denotes the subtangent with respect to Ox. For the modern reader, the process may seem a bit fast[4]. In fact Tschirnhaus will explain in another contemporary letter to Leibniz (A III 3, No. 357) the full details of his method (on the same sample circle), which makes no use of partial derivatives, but uses Descartes' designs and Viète's law of homogeneity.

Tschirnhaus and the tangents to an *arbitrary* transcendent curve

The above process also remains interesting by the analogy of treatment which Tschirnhaus proposes for the non-algebraic curves, curiously, from then on, called by him 'mechanical' (and not transcendent) – this is an explicit reference to Descartes. As for these, Tschirnhaus simply *assimilates* them to the algebraic ones,

---

4   Of course, Tschirnhaus does not use partial derivatives. By his method, he finds (p. 490) the
value $\sigma = \dfrac{2ax - x^2}{a - x}$.

while continuing to define them (p. 490) by polynomials in two variables – these variables, however, need no longer be coordinates, but, possibly also, lengths of arcs. Tschirnhaus then explains that, for him, the calculation of the quotient Z is done in exactly the same way. This definition, which is not followed by any example in the text, is simply the object of a brief incidental remark. Limited as it can be (it does not obviously describe all the transcendent functions!), it was nevertheless, to my knowledge, new at the time, and could be fruitful: it allows, for example, to consider the general equations[5] of curves as F (x, Arc sin y) =0 or F (x, Arc tan y) =0, where F is a polynomial in two variables, and the tangents to such curves. Here emerges the procedure which would become standard; that of the 'normal vector' for the calculation of the tangents to an implicitly defined curve, which had been brought to light by Huygens (see chap. XII) and that we detail below in Chapter XV.

XI-B THE RECEPTION BY CRAIG (1685 AND 1693)

John Craig and the impossibility of the quadrature of the transcendent figures

John Craig was closely interested in the calculation of quadratures and published in England, over a period of eight years, two volumes on the subject; at first, the *Methodus figurarum lineis rectis et curvis comprehensarum quadraturas determinandi*, ([Craig 1685]), and then the *Tractatus mathematicus de figurarum curvilinearum quadraturis*, ([Craig 1693])[6]. Craig, who was one of the first to introduce the mathematical works of Leibniz to England, willingly quotes this one and his differential calculus (cf. for example the introduction to the *Tractatus*, page 1). From 1685, (when he was twenty two years old), he had already made reference to the 'transcendent curves', a term which was therefore directly imported from Leibniz. So, in the *Methodus*, he evokes the question of the transcendent by this reflection, located in the traditional context of quadratures:

> And already nothing is missing so that the method which I described to determine the quadrature of figures extends in all the figures (with the exception of those that are limited by transcendent curves, for which no known method applies so far), except that I must get rid of two difficulties which can occur in some cases (…) ([Craig 1685]), 26).

In reality, despite his declared interest for the methods of Leibniz, Craig never ceased to express strong objections regarding the treatment of the transcendent curves, whatever the calculation, differential or integral. In 1685, as we have in fact just seen, he refused, as regards quadratures, to apply his method to the transcendent curves – it was a matter of integral calculus there. However, his reluctance extended in the same way to the differential calculus. Indeed, as L'Hospital underlines clearly, (cf. *infra*

---

5   The idea maybe came to Tschirnhaus by the Arithmetical Quadrature of the Circle of Leibniz, where this one manipulates a relationship between x and Arc tan x.

6   Our references come from [Breger 1986], 131. On Craig's mathematical work, see [De Morgan 1852].

section XI-D, in this same chapter[7]), he also refused, in his *Tractatus* of 1693, to extend to the transcendent curves the differential method for determining tangents.

XI-C THE RECEPTION BY STURM (1689)

John-Christopher Sturm and the 'transcendent degree'

Sturm published, in 1689 in Nuremberg, his *Mathesis enucleata* (…), the title of which means approximately: *On Mathematics concentrated in its most interesting* (…)[8]. This book, though dated 1689, admittedly introduces the term 'transcendent'; yet it reflects, in fact, a rather short reflection on behalf of the author on the issue in question.

In the proposition XLIII (page 181), Sturm claims at first to demonstrate the irrationality of π, based on Leibniz's theorem on the Arithmetic Quadrature of the Circle which he explicitly quotes (see page 181, *Scholium*). Breger correctly writes that Sturm's approach is fully vain[9]. Further on (page 321), at the conclusion of his chapter V, citing these 'very important' (*insignissimis*) authors that are Leibniz and Craig, Sturm evokes in a very vague way, the "indefinite degree" and the "transcendent degree":

> We are now assured by very important mathematicians, Leibniz and Craig, that the curves of this kind, even if we cannot express them by equations of a known degree, fall under the Geometric ones, even if this is in spite of Descartes, because they admit equations which are – at least – of an indefinite degree or still a transcendent degree, and can be submitted in the same way to the calculation and to its consequences; it may be of another nature than that we usually use.

As we see, the text remains vague regarding the equations of 'a transcendent degree'! Sturm takes refuge behind the authority of Leibniz and Craig – however, as we have just seen, the latter had considered that no method of quadrature was known for the transcendent curves. As regards the actual contents of the transcendence, Sturm doubtless thinks, like the young Leibniz, of equations with letteralized exponentials, which would be the only model. Further on, he will give two references in the texts of Leibniz in the *Acta,* firstly from 1684 – page 234 (*i. e., Nova Methodus* (…)) – and secondly from 1686 –pages 292 and 294 (*i. e., De geometria recondita* (…)).

---

7   See also in V-E above, the critique of Craig by John Bernoulli.
8   The full title is *Mathesis enucleata, Cujus Praecipua Contenta Sub Finem Praefationis, uno quasi obtutu spectanda exhibentur.*
9   [Breger 1986], 132.

XI-D L'HOSPITAL, DIFFERENTIAL CALCULUS
AND TRANSCENDENT CURVES

The *'Analyse des infiniment petits'* (1696)

Breger naturally quotes here the *Analyse des infiniment petits pour l'intelligence des lignes courbes (Infinitesimal calculus with applications to curved lines)*[10], the well-known work of the Marquess of L'Hôpital. This text of 1696 is important in the history of mathematical ideas, because it remained for a long time – roughly, up to the treaties of Euler in the 1750s – the only published treaty on the differential calculus. We also know that the book is deliberately only concerned with differential calculus, as the author explains in his preface. The Leibnizian calculus is composed of two 'symmetric' parts: the differential calculus and the integral calculus. Leibniz had initially declared his intention to publish on the integral calculus. The Hospital wanted to leave him this glory and thus published on the differential calculus only. The term 'transcendent' is really only used rarely in the text; mainly in the preface and the last lines of conclusion.

The preface falls within the framework of a legitimate criticism of Craig, who had refused in his *Tractatus,* published three years earlier (1693), to recognize that the calculation of tangents (which is indeed differential) applied equally to transcendent curves. Exposing his plan of the book, l'Hôpital writes:

> I divide it into ten sections. The first one contains the principles of the Calculation of the differences. The second shows how we have to use it to find tangents of all kinds of curves, whatever is the number of indeterminates in the equation which expresses them, although Mr. Craig did not believe that it was able to be extended up to the mechanical or transcendent curves.

The very term 'transcendent' thus hardly appears in the book but many explicit instances of transcendent curves are present. In the second section, L'Hôpital examines more particularly the construction of the tangent at the running point of diverse curves. Also, his method is strictly differential: it is not necessary for him to specify the equations of such curves, but only to describe (in symbolism) a geometrical construction of them, and then to differentiate the writings obtained by a differential method of 'small triangles'.

The organization of the book

We cannot analyse the book here in detail, but we will briefly highlight its remarkable design. L'Hôpital proceeds each time by going from the general to the particular; he starts by stating and solving a general 'problem' of the determination of the tangent to a curve, the equation of which is left unsettled; then, in an 'example', he particularizes the data to obtain a specific curve of the considered example. It should be emphasized that the curves are more usually given here by a characteristic property – differential or metric – rather than through an explicit equation.

10   [Breger 1986], 132.

Accordingly, proposition VI (p. 21, § 24, figure 11) describes a general geometrical problem concerning (among other things) – 1 °) An arbitrary given curve, this definition thus potentially includes the case of transcendent curves, and – 2 °) a metric relation – equally arbitrary and given – involving two different numbers, measuring lengths. From these two data, the author deduces, as a consequence, the construction of some other curve, called 'resultant', and explains how to find the tangent at the running point of this one[11]. Then, in an 'example' immediately following (§ 25), he particularizes the given curve and the relation, and examines the curve obtained: in the case quoted in § 21, the resultant curve is the conchoid of Nicomedes (page 22); thus he proves that he knows how to find the tangent to any point on it[12]. He does the same at Proposition VIII (page 23, § 27, figure 13) which leads for 'example' (p. 24, § 28, figure 14), to the construction of the tangent at the running point of the cissoid of Diocles.

L'Hôpital and the tangents to the transcendent curves

Yet cissoid and conchoid are both algebraic curves – cubic for the one, quartic for the other. However, his method also serves the author for the transcendent curves; to define them as well as to construct their tangents. Under these conditions, however, Proposition XII (p. 34, § 37, figure 26) which follows, describes – as we detailed it above – a very general geometrical problem, concerning two arbitrary given curves and a metric relation equally given. From these data, the author deduces (by particularizing) the construction of a new other curve, which he defines in this case as the *logarithmic curve* (for us now, it is the exponential curve). Hence, the logarithmic curve is naturally and simply characterized (page 35) in that its subtangent is constant. One finds in the same way another clear example; that of the logarithmic spiral, in section VII, dedicated to caustics by reflexion (§ 140, p. 127).

L'Hôpital also takes into account various curves that are *genuinely transcendent;* however, without using the term. These will be mainly four in number; the spiral of Archimedes[13], the logarithmic curve (for us, the exponential)[14], the logarithmic spiral[15], and the roulette (for us today, the cycloid)[16]).

---

11   He so finds (page 21) the general formula $FT = \dfrac{sy^2 dx}{x^2 dy}$ and he concludes "we shall then finish

the rest by means of the difference of the proposed equation [i. e. the relation]".

12   Page 22.

13   Page 20.

14   Pages 35 and 88.

15   Pages 36, 89, 117 and 127.

16   There are many references to the roulette in the book. For example, pages 43, 64, 91, 114, 115, 116, 151 and 156.

The glory of Leibniz

In the last lines of the work (p. 181), L'Hôpital returns to general considerations on the transcendent curves; they are for him the elements of a song of glory at the address of Leibniz. Fully in line with the constant claim of the latter, L'Hôpital insists, *differentiating is always easier*, in particular in the case of the radicals:

> We clearly see by what we have just explained in this section, how we have to use the method of Mrs Descartes and Hudde to solve these kinds of questions, when the curves are geometrical. But we also see at the same time that it is not comparable to that of Mr Leibniz, which I tried to explain completely in this Treaty, because the latter gives general resolutions where the other provides only specific ones; it also extends to transcendent lines, and it is not necessary to remove the incommensurables; what would often be very impracticable.

XI-E JOHN BERNOULLI. ABOUT THE ORGANISATION OF THE TRANSCENDENT COMPLEXITY (1695–1730)

Bernoulli, Huygens and Leibniz, between hypotranscendence and hypertranscendence

Transcendence issues were the subject, from 1695, of a voluminous correspondence from John Bernoulli to Leibniz, upon which we have already partly commented above in V-E, *Percurrence and transcendence, Leibniz and John Bernoulli – the story of a misunderstanding*. There, one discovers to what extent the reception by Bernoulli, of the term and the concept of transcendence, was attentive.

Whilst the direct colleagues or correspondents of Leibniz cited above (Sturm, Craig, Tschirnhaus, Huygens, L'Hôpital), were often satisfied with vague considerations on the nature of transcendent entities, Bernoulli, instead, will focus on a relevant attempt of construction, of analysis and inventory. So, for example, he was perfectly entitled to laugh at Huygens, so much for the terminology of 'hypertranscendence' of the latter, as for the insufficiency of his reflection on the subject. In *Principia Calculi Exponentialum, seu percurrentium*[17] of 1697, Bernoulli indeed wrote:

> I had thus conceived the exponential quantity as a middle term between the algebraic and transcendent: it approaches the algebraic, this one being composed of finite terms, namely indeterminate; but for the transcendent, we cannot show any algebraic construction. Therefore we must consider how much was inappropriate the joke of Huygens, when he called *hypertranscendent* the exponential curves[18]; with more seriousness and relevance, he would have had to call them *hypotranscendent*. The logarithmic curve itself indeed belongs to this kind of curves (the simplest certainly, of all the transcendent). And, from the latter, all other exponentials can be constructed, and their tangents determined; as is evident from what has been already said.

---

17   *Œuvres*, Bd. 1, 180.

18   This is a reference to the terms of the letter from Huygens to Leibniz on 18th November, 1690 (GM II, 56), as discussed in detail below in chapter XII. The exact term used by Huygens is '*supertranscendent*'.

This irony is quite relevant. It is indeed clear that for anyone trying to explore the *terra incognita* of the transcendence, the exponentials constitute one of the most easily affordable areas, both the simplest and the best known symbolically. It is thus right that they are, for Bernoulli, somehow transcendent to a lesser degree, 'hypotranscendent' thus, and not 'hypertranscendent', as wrote Huygens.

Bernoulli and the "degrees of transcendence"

Trying to capture the complexity of an unknown domain by the organisation of degrees in this complexity is an epistemological approach that Bernoulli greatly appreciated. We have already noted above that he had implemented it in the framework of the degrees of percurrence in the above text of March 1697[19]. Later, he will resume the same approach in the larger context of the *degrees of transcendence*.

We shall quote in connection, always following Breger[20], a later text (subsequent to the death of Leibniz) published in *Acta*, dated August, 1724, *Convenient and natural method to reduce the quadratures of the transcendent curves of any degree to rectifications of algebraic curves*[21].

The article indeed contains numerous references to the transcendence! According to its title, Bernoulli proposes to demonstrate that the issue of the calculation of the *quadrature* of a transcendent curve of any degree is equivalent to that of the *rectification* of a certain algebraic curve. In other words, he suggests establishing something like this: the areas (of the transcendent curves) and the lengths (of the algebraic ones) are quantities of the same species[22]. The problem for him is to define clearly what is *"a transcendent curve of any degree"*? Here we are again at the heart of the inventory issue.

For Bernoulli, the first (and fundamental) model of a transcendent curve is that obtained as the quadratrix of an algebraic curve, which is not itself algebraic[23]. It is a conception of the transcendence on which he will not vary. We exposed above (cf. chap. VI) the idea that he had, still in 1730, of the transcendent functions[24]. For him, they were necessarily associated with quadratures, as follows; given an arbitrary function p, by putting before it, indiscriminately, a sign of integration, to obtain $\int p\,dx$, we always get a transcendent function, namely $F = \int p\,dx$. In the cited article from 1724, Bernoulli then resumed and generalised these conceptions.

---

19 See *supra* 'On a possible hierarchy in percurrence', in Chapt. V-E.
20 [Breger 1986], 132.
21 *Methodus commodo & naturalis reducendi Quadraturas transcendentes cujusvis gradus ad Longitudines Curvarum algebraicarum = Oeuvres*, Bd. 2, 591.
22 In modern terms, it is a question, in essence, of showing that every integral of the form
    $\int f(x)\,dx$ (with f(x) ≥ 1) can be written in the form $\int \sqrt{1+(g')^2(x)}\,dx$.
23 In modern terms, it is a primitive of a rational fraction that is not itself a rational fraction. This case is usual (see IX-D).
24 See John Bernoulli, 'Méthode pour trouver des Tautochrones, dans des milieux résistants, comme le carré des vitesses' = *Œuvres*, vol. 3,174.

On the comparative transcendence orders for the quadratures

In the initial paragraphs 5 to 8 (p. 583–586), Bernoulli wants to show that the quadrature of any *algebraic* function is generally equivalent to a certain rectification. In a second step, however, he must deal with the case of the quadrature of a transcendent function. The crucial point for us here is in page 591 (§ XV and XVI). It lies in the definition of 'degrees of transcendence'. These are simply obtained by the successive degrees of repeated integrations (sum of sum, sum of sum of sum, etc.). Bernoulli so writes (Chapter XV, p. 591):

> I come now to the transcendent areas of higher degree (…). Let A, B, C, D, E be a sequence of curves, constructed on a same abscissa x, such that each is so generated by the following one: the area of the curve A is proportional to the ordinates of a curve B[25], and the area of B is proportional to the ordinates of a curve C, and then, the area of this curve C is proportional to the ordinates of a curve D, and so on (…) each one is thus constructed by the quadrature of the one which precedes it immediately. If it is now assumed that A is algebraic, it will have for ordinate an arbitrary function of the abscissa x, say p; let us suppose nevertheless that the elementary area p dx is not integrable; I shall say that the quadrature of the first curve A is *transcendent of the first order*; that the quadrature of the second curve B is *transcendent of the second order*; the same applies to the third C, to the fourth D, to the fifth E, etc. which are *transcendent of the third, fourth, fifth order*, etc.

This scheme thus organized a natural scale of comparison in the transcendence of areas. In view of Bernoulli's perspective on transcendence, described above, the approach was obviously completely justifiable and appears relevant. Bernoulli, however, wants to demonstrate that, for all the degrees of transcendence so defined, the quadratures are equivalent to a single rectification. Chapter XVI immediately following then develops calculations associated with the previous scheme and obviously concerns what are called multiple integrals today. Bernoulli first resumes the previous notations and writes (p. 592) the following relations (where p is an algebraic function):

"First area $A = \int p\,dx$

Second area $B = \int dx \int p\,dx$

Third area $C = \int dx \int dx \int p\,dx$

Fourth area $D = \int dx \int dx \int dx \int p\,dx$

and so on, up to infinity".

---

25   That is to say, B is a quadratrix of A.

Only the first order subsists *in fine*

On these values of areas, he then performs calculations of integration by parts. He obtains the results below (p. 592):

$$
\begin{aligned}
&\mathbf{1}\ldots\int p\,dx = \int p\,dx\\
&\mathbf{2}\ldots\int dx \int p\,dx = x\int p\,dx - \int p\,x\,dx\\
&\mathbf{3}\ldots\int dx \int dx \int p\,x = (xx\int p\,dx - 2\,x\int p\,x\,dx + \int p\,xx\,dx):\mathbf{1.\,2}\\
&\mathbf{4}\ldots\int dx \int dx \int dx \int p\,dx = (x^3\int p\,dx - 3x^2\int p\,x\,dx + 3x\int p\,x^2\,dx - \int p\,x^3\,dx):\mathbf{1.\,2.\,3}\\
&\qquad\qquad\qquad\qquad\qquad\mathbf{\&c.}
\end{aligned}
$$

To explain and discuss these results for the reader of today, we prefer to resume the calculations in modern notation[26].

Relation 1 is a truism. To examine relation 2, we define:

$$
J(x) = \int_0^x du\left(\int_0^u p(t)\,dt\right)
$$

John Bernoulli then integrates by parts[27]:

$$
J(x) = \int_0^x \alpha\,d\beta = \left[\alpha\beta\right]_0^x - \int_0^x \beta\,d\alpha\,, \text{ with } d\,\beta = du;\ \beta = u, \text{ and}
$$

$$
\alpha = \int_0^u p(t)\,dt\,, \text{ hence } d\alpha = p(u)\,du. \text{ Therefore:}
$$

$$
J = \left[u\int_0^u p(t)\,dt\right]_{u=0}^{u=x} - \int_0^x u\cdot p(u)\,du = x\int_0^x p(t)\,dt - \int_0^x u\cdot p(u)\,du
$$

This he writes $x\int p\,dx - \int p\cdot x\cdot dx$ (this is relation 2 *supra*).

As such, any quadrature of transcendents of degree two (namely J(x), area B) – thus having its origin in a double summation – is reduced, in terms of its complexity, to the degree one (that is, a simple summation). The same applies for a triple summation (area C). If we indeed define with modern notations:

$$
K(x) = \int_0^x dx\left(\int_0^u du\left(\int_0^t p(t)\,dt\right)\right)
$$

Then one has $K(x) = \int_0^x J(u)\,du$. In this formula, J(u) is given by the above relation:

---

26   The modern notation indeed distinguishes, rightly, between the symbolism of the variable of integration, which is 'mute', and that of the 'limits' on the integral.

27   As regards the integration by parts, Bernoulli makes a reference (p. 592) to one of his articles which appeared in *Acta* of 1694.

$$J(u) = u \int_0^u p(t)\,dt - \int_0^u t \cdot p(t)\,dt$$

One thus gets:

$$K(x) = \int_0^x du \left(u \int_0^u p(t)\,dt \right) - \int_0^x du \left(\int_0^u t \cdot p(t)\,dt \right) = K_1(x) - K_2(x)$$

Integrating by parts $K_1(x)$ and $K_2(x)$ separately, we have:

$$K(x) = \frac{1}{2}\left[x^2 \int_0^x p(t)\,dt - 2x \int_0^x t \cdot p(t)\,dt + \int_0^x t^2 p(t)\,dt \right]$$

This is the result 3 *supra* that Bernoulli provides for area C (page 592); once again, it thus involves simple integrals only. He also gives the result 4 for area D and, he says, we get an analogous form for any order of summation n. He provides an overview of the general formula.

Whatever n, all the summations of order n are thus computed by summations of order one. Accordingly, whatever the order of transcendence of the curve (i.e. the order of the summation), Bernoulli, by successive integrations by parts, brings back its quadrature to simple summations; namely, linear combination of quadratures of algebraic functions. Each of these is for him equivalent to the rectification of an algebraic curve; thus, this completes the proof of the announced result.

Bernoulli and the *repetition* of the transcendent creation

Our conclusions in this chapter join those of V-E, by underlining that the design of the transcendence in Bernoulli has assumed a double face: the letteralization of the exponentials on one hand, the summation on the other hand. Each of them produces a form of transcendence; the first being called 'percurrent' by Bernoulli. Thus, it is interesting to note that Bernoulli will always be led by the same strong idea; namely, to organize a hierarchy in the realm of what he identified as transcendent. This approach is quite understandable: what could be more natural, in order to domesticate the unknown, than to try to organise it in classes? To do this, the approach of Bernoulli will be invariable, the same in both cases; namely, *to repeat the operation which created the transcendence.* That is to say, take the exponential of an exponential in the percurrent case (thus creating an exponential of degree 2), or calculate the quadrature of a quadrature in the case of the transcendent areas (areas of order 2). Such a process is also very natural. Thus, Bernoulli obtained, on one hand, a hierarchy of 'percurrent expressions' (cf. V-E) and, on the other hand, the hierarchy of orders of the transcendent areas that we have just explained. Leibniz, it seems to me, did not feel concerned by this epistemologically probative value of the repetition of a creative operation.

Still on the epistemological level, however, we will have to observe as a conclusion of the preceding section that, contrary to his concept of the *degree of per-*

currence (see V-E), that of the *degree of transcendence* has been emptied of its substance by the author.

XI-F ON THE PHILOSOPHY OF SIMPLICITY.
DESCARTES *VERSUS* NEWTON

The 'geometrically irrational' curves in Newton (1687)

Following the references of Breger about Newton[28], we first examined the lemma XXVIII, in the *Principia* of 1687[29]. As a consistent geometer of his time, Newton is only interested in one thing (pages 116–117): the number of points of intersection of a plane curve with a straight line. In the case of the generation of some specific spiral – an example that he develops in detail – such a number is clearly infinite, and the curve cannot thus be the object of an equation of finite degree. He insists on this point a lot, by multiplying examples of counting the number of intersection points of two plane curves and, as a corollary, the degree of the equation that provides the latter[30]. He concludes in a corollary (p. 118) by a typology between *geometrically rational and geometrically irrational* curves:

> I call geometrically rational curves, those for which the relation between abscissas and ordinates can be determined by equations in finite terms. Other curves, such as spirals, quadratrices, trochoids, etc., I called them geometrically irrational curves.

Thus, in this text, Newton does not take the Leibnizian terminology that distinguished between algebraic and transcendent curves. Nor does he use Descartes' categorisation: geometrical curves/mechanical curves. This point is rightly noted by Breger[31]:

> Newton introduces this terminology in a rather common way and he does not discuss its relation with the Cartesian mathematics, nor the question of the legitimacy of the non-algebraic curves.

On the other hand, the emphasis of his terminology on the adverb 'geometrically' obviously implies that, for him, all curves are, in the first place, *geometrical objects*. However, it will be necessary for him to appeal to the *symbolical* qualifiers 'rational' and 'irrational' for categorizing them! Those he called *geometrically ra-*

---

28 Breger, 1986, p. 131, n. 77.
29 Book I, section VI, corollary to Lemma 28. We will use and reference the French translation of Émilie du Chatelet = [Newton 1759]. Lemma 28 was the subject of specific studies. See [Pesic 2001], 215. It is also commented upon in [Guicciardini 2005], 76. See also [Peiffer 1989] and [De Gandt 1986].
30 So Newton explains in detail (pages 116–117) 1°) that the intersection of two conics generally leads to an equation of the fourth degree. 2°) that the intersection of a conic and a cubic generally leads to an equation of the sixth degree. In this Lemma, Newton was actually interested in some issue about the kinematic generation of a spiral, based on a closed curve of the plan (an 'oval'), and associated this with what is now called the 'law of areas' (see [Guicciardini 2005], 76).
31 [Breger 1986], p. 131, n. 77.

tional curves are *algebraic* curves in Leibniz, or *geometrical* in Descartes. For Newton, they are indeed described 'by equations in finite terms'; that is to say, by polynomials. The other curves (and the categorization is here also, quite naturally, a binary one) are *geometrically irrational*. Let us note, however, that this term is a bit curious, as it concerns objects which are essentially geometrical.

Newton gives three examples of geometrically irrational curves, (spirals, quadratrices[32], trochoids), which may actually be cut by a straight line in an unbounded number of points. One can also think that he includes into this category the curves the equation of which involves the sum of a power series (it's a mathematical field in which he excelled) – which implies an infinite number of terms: this is the case of the usual trigonometric curves, of equations $x \rightarrow \sin x$, or $x \rightarrow \cos x$. These two curves, however, may also be cut by some straight lines in a finite number of points. We shall note that on the other hand, in this text, Newton does not classify usual curves which, as the exponential or the arctangent, while, being irrational (in the Newtonian sense), are cut by any straight line in a finite number of points. Contrary to Leibniz, Newton makes no effort in this text towards an attempt of a reasonable inventory of the non-Cartesian curves (i. e., for him, geometrically irrational).

Anyway, it is the terminology of transcendence of Leibniz that prevailed; that of Newton did not survive, as Breger points out pertinently[33].

From Descartes (1637) to Newton (1707).
Symbolical simplicity *versus* geometrical simplicity

Twenty years after the *Principia*, the issue of a classification of curves (and that, correlative, of their legitimacy) found at Newton's a new occurrence in an interesting extended appendix to the *Arithmetica Universalis* of 1707, the *Aequationum Constructio linearis*[34].

Regarding the curves, this important text clearly explains Newton's mathematical ontology – basically geometrical; for him, mathematical existence is coextensive with the faculty (and with the simplicity) of construction. Newton begins by recalling Pappus' classification of geometrical problems (planar, solid and linear). Next, he writes (p. 212) that "it was discovered recently" that one has to accept all the curves in geometry, provided they are equipped with an equation. If the reference to the mathematical work of Descartes is obvious, the name of Descartes is not mentioned! Such a prescription would allow, according to Newton, the classification of curves in kinds, according to a hierarchy of simplicity – again, a conception indisputably Cartesian. However, Newton continues (p. 213): "the simplicity is not that of the equation, but that of the facility of construction of the curve". For New-

---

32   This expression must be understood as applying to the quadratrix of Hippias, or Dinostratus. Its Cartesian equation has the form $x = y \cotan (y/a)$.

33   "As for the name for the differential calculus and the integral calculus, here also, the terminology of Leibniz has prevailed over that of Newton": [Breger 1986], p. 131, n. 78.

34   [Newton 1732], 212–241. See [Breger 1986], p. 131, n. 78.

ton, the rule was thus absolute, not to construct a curve (he says *lineam*), by means of curves of a higher kind than necessary (pages 212–213). Here we believe we hear Descartes himself, on this issue of simplicity, when he wrote in his *Géométrie* (ATVI, 444):

> On the other hand, it would be a fault to try vainly to construct a problem by means of a class of lines simpler than its nature allows.

This conclusion is hasty, however. Indeed, let us return briefly to Descartes to compare his position with that of Newton on this issue of simplicity[35]. Firstly, we recall that after many vicissitudes, the figure of thought of simplicity had finally found its real place and its definitive dimension in the *Géométrie*; that is, that of a real organizing criterion of the research. As such, it fell again within the problematic of the Cartesian order: as a corollary of the aforesaid dialectic of the *'fault'*, there will indeed be, in the conceptions of Descartes, a single *'order of simplicity'* a minima and, as a new corollary, a single method (of construction) will thus be acceptable.

However, for Descartes, this order of simplicity was necessarily symbolic, embodied by *the degree of the equation* – a concept that he himself had only recently introduced in mathematics[36]. Descartes thus hoped to establish a total order between the simplicities of construction, in accordance with the constant principles of the Cartesian epistemology[37].

As to Newton, he argues in detail – farther in this same *Aequationum Constructio* – about such a requirement, by discussing the comparative simplicities of the circle and the parabola: the circle is simpler to construct than the parabola, but the latter has an equation simpler than the circle. Thus, the circle for him is intrinsically simpler[38]. For Newton, who appears here as a pure geometer, algebraic expressions do not contribute to the simplicity of the geometrical constructions and, in every case, the simplicity of the construction is essential (p. 213). Also, he insists on such an exclusion of the symbolism (p. 214):

> The equations are expressions coming from the arithmetical calculation, and they have no real place in geometry.

Such a conclusion is exactly contrary to that of Descartes and his considerations on the degree! To conclude, it will thus be necessary for Descartes, just like Newton, to fulfill an obligation of some specific kind of process, namely *"a minima"* simplicity, but it will not be the same simplicity for our two geometers.

---

35 See 'Le Problème de Pappus. Hiérarchie de simplicité des courbes et critère algébrique', in [Serfati 2008b], 27–34.

36 Recall that, in fact, Descartes defines not exactly the degree, but the 'genre': degrees grouped in pairs according to (2n–1,2n), when you define the genre n.

37 On the primacy of order in the Cartesian epistemology, see [Serfati, 1994], 71–72. On the hierarchy of complexity for the procedures of the construction of curves, see [Serfati 2008b], 33.

38 In a suitable coordinate system, the equation of a circle is $x^2+y^2 = R^2$, containing two squares (those of the abscissa and ordinate). It is symbolically more complex than that of the parabola, which is $y^2= 2px$, with only one square (R and p are parameters). This example is paradigmatic of the difference between symbolic simplicity and simplicity of construction.

Further on (p. 214), Newton returns to trochoids, by repeating his claims of the *Principia*: and states that it would be absurd not to admit the trochoids in geometry, this would complicate the description of the problems.

A trochoid is indeed easy to describe in terms of its geometrical construction: it is the curve described in a plane by a fixed point on a circle as the latter rolls along a straight line – the cycloid (or 'roulette') is a special case. On the other hand, the formation of the equation of this geometric locus – a relatively complex operation for the time of Newton – leads naturally to a *parameterization*, rather than a standard *implicit representation* (F $(x, y) = 0$ – see below Chapter XV).

CHAPTER XII
THE TRANSCENDENCE BONE OF CONTENTION BETWEEN HUYGENS AND LEIBNIZ (1690)

XII-A THE CONTROVERSY

Leibniz and Huygens

The reception of the notion of transcendence by Christian Huygens, of course, was particularly important to Leibniz. I examine in this chapter a controversy between both scholars contained in an exchange of letters dated from the second half of 1690 (August to November), from which I retain five main letters[1]. I will first briefly recall some historical elements on Huygens' attitude with respect to Leibniz's calculus.

After the initial occasion of their Parisian exchanges, when the young Leibniz was a follower of Huygens, the correspondence between both men was suspended for a few years. This period was also the time when Leibniz gradually set up the full architecture of his differential calculus. But Huygens, fully convinced by the ancient geometrical methods, was strongly resistant to it. As an example, he believed it was not necessary for him to read Leibniz's articles in the *Acta*.

The correspondence discussed here takes place at the beginning of a second period of exchanges between Leibniz (writing from Hanover) and Huygens (most often responding from The Hague). Huygens is still widely skeptical about the virtues of Leibniz's infinitesimal calculus, claiming that he himself had previously reached the same results – by using geometrical methods which he considered classical. On the other hand, Leibniz, broadly supported by the practical and theoretical successes obtained *via* his calculus, was much more self-confident. Then, his legitimate objective was to convince Huygens, to convert him to the new calculus. The present chapter also aims to describe this attempt at the conversion of Huygens to the notion of transcendence.

Let us now return to the very interesting quarrel of 1690. The detail is mathematically quite complex, and I develop in this first part (XII-A) only the main ideas and major articulations. The second part of this chapter (XII-B) is devoted to a more detailed mathematical commentary.

This controversy is actually emblematic – it opposed an old-school geometer (Huygens), an ardent defender of an algebraic curve (a cubic one) to a 'modern' analyst (Leibniz), champion of a transcendent curve (with a *letteralized* exponential). Moreover, as we shall see, the dispute was exacerbated by a discrepancy between the two protagonists concerning a fundamental definition, that of the subtangent to a curve.

---

1   Breger is also interested in this episode. See ([Breger 1986], 132).

The controversy of 1690 between Huygens and Leibniz

The origin of the dispute lies in the first letter from Huygens to Leibniz, 24th August 1690[2], in which Huygens addresses an enigma, as a challenge to test the new infinitesimal methods: he mentions, without any specific details a problem of plane (geometrical) locus, which is classical for him, (that is to say, the problem was not differential). He does not communicate to Leibniz the method he used to solve it, nor the equation of the curve he found. Without calculus, however, he exhibits the value of the sub-tangent of the latter, namely:

$$BD = \frac{2x^2y - a^2x}{3a^2 - 2xy} \quad (3)$$

He then asks Leibniz (p. 46) to use his infinitesimal methods to find the curve in question, starting from the value of the sub-tangent. So settled, it was for Leibniz a classical problem of the inverse method of tangents, typically requiring his differential calculus (see above, Chapter VI). In his reply dated 3rd (13th) October 1690, dated from Hanover[3], Leibniz, who does not supply more calculus, then declares (p. 621) that he found as a solution the transcendent curve with equation[4]:

$$\frac{x^3y}{h} = e^{\frac{2x \cdot y}{a^2}} \quad (4)$$

He argues:

> It is a transcendent equation where the unknowns enter the exponent (…) In the *Acts of Leipzig*, I happened to mention these equations with unknown Exponents, and when I can obtain them, I prefer to those exclusively formed by means of sums and differences. Thereby, they can always be reduced to differential equations, but not vice versa.

One recognizes in this comment a fundamental source of Leibniz's interest in exponential equations: they are immediately amenable to his differential calculus which, contrary to the integral calculus, does not raise any problem of application (see on this point the analysis above, IV-C5).

The third letter[5] is also from Leibniz. He now wants to justify his result (4) to his interlocutor, so he uses a reciprocal method: by differentiating relation (4), he says (p. 53), one finds (3). However, as we shall see in the second part of this chapter, the formula he uses for the value of the sub-tangent is not the same as Huygens': it is, in fact, the opposed one! This is not necessarily a mistake: the definition of the sub-tangent is indeed conventional. It is not a length but a number (an algebraic value), and one can choose for definition either $\overline{HT}$ or $\overline{TH}$ (see XII-B for the mathematical details). However, this ambiguity will be at the origin of the controversy.

---

2   AIII, 4, N. 271, p. 545–550.
3   AIII, 4, N. 283, p. 619–624.
4   In formula (4), Leibniz uses the sign *b* instead of *e*.
5   AIII, 4, N. 287, p. 640–647. It is dated October 28th, 1690.

The fourth letter is Huygens' reply[6], from The Hague, dated November 18th, 1690. This time, Huygens gives Leibniz the complete details of the geometrical problem, which was at the origin of the question. At the same time, he supplies Leibniz with the solution he himself found for it (p. 56), namely the algebraic curve of degree three (a cubic one), with this (quite homogeneous) equation:

$$x^3 + xy^2 = a^2y \text{ (relation (1))}$$

This is an algebraic solution, which is obviously quite incompatible with Leibniz's transcendent result! Huygens then develops the skepticism of the "classical geometer," and does not lack for irony:

> Thus it is not one of these Transcendents, as suggested by your equation (…) As for me, I confess that the nature of these super-transcendent lines, where the unknowns enter the Exponent, seems so obscure, that I would not think introducing them into geometry, unless you notice some significant utility.

Huygens then mentions what he calls Leibniz's "speculations" about the catenary. As to the content of the reply, it is clear that Huygens is wrong on a theoretical level (*letteralized* exponentials are certainly a fully interesting mathematical expression).

In his response of 14th (24th) November 1690[7], Leibniz (p. 666) dedicates himself to computations: if the equation of the curve is (1), that is to say, the cubic $x^3+xy^2=a^2y$, as proposed by Huygens, then, he says that he himself finds for the sub-tangent:

$$\frac{a^2x - 2x^2y}{3a^2 - 2xy} \text{ and not "} \frac{2x^2y - a^2x}{3a^2 - 2xy} \text{ , as you suggested me."}$$

In other words, he finds again $- (3)$ and not (3). This result is, however, natural for Leibniz, considering his choice for the formula of the sub-tangent (one can find in XII-B a more complete mathematical analysis of the question).

The two results are actually opposed, and Leibniz then rightly insists that they are not compatible. As such, he decides to resume the entire question (p. 667), this time by a differential calculus, that is by differentiating both of the previous equations (4) and (1). He finds, for his solution ((4), transcendent):

$$3\frac{dx}{x} + \frac{dy}{y} = 2xdy + 2ydx \text{ (relation 5)}$$

As to this relation, he says, "I can solve it, because the sum of 2xdy+2ydx is 2xy." Meanwhile, in the other hypothesis (Huygens' algebraic issue), Leibniz finds:

$$3\frac{dx}{x} - \frac{dy}{y} = -2xdy + 2ydx \text{ (relation 6)}$$

6   AIII, 4, N. 291, p. 654–658.
7   AIII, 4, N. 292, p. 659–669.

In this case, he says, one does not simply reach the solution:

> "Here it is different, and one must use other skills, which I had not used, because I reached very easily what you asked me."

Both numerical results are exact and, on this theoretical aspect of the differential calculus, Leibniz is obviously right: the 'Leibnizian' relation (5) is an exact differential, which therefore can be immediately integrated, contrary to Huygens' formula (6) (see XII-B). Leibniz's argument is apparently correct. His results and those from Huygens are, however, reversed, and thus seem incompatible, as a result of their different choices for the value of the sub-tangent!

That was the first (and difficult!) contact between Leibniz and Huygens concerning transcendence. As we have seen, Huygens persisted in his refusal to take into account Leibniz's *letteralized* exponentials as legitimate mathematical entities. However, as we have also seen, the mathematical context of this controversy was genuinely difficult and complex, especially insofar as a complete opposition between both interpretations perturbed the discussion.

With this cubic-exponential example, Leibniz did not succeed in convincing Huygens of the central mathematical importance of *letteralized* exponentials and of transcendence. However, he did not give up, and in resuming his effort at conversion, he began to develop a much simpler example in the fifth letter (AIII, 4, N. 292, p. 659–669) from which he continued to argue two months later, in his correspondence dated January 27th, 1691 (AIII, 5A, N.6, p. 38–51).

A lesson on method from Leibniz to Huygens: an equivalence between power series and exponentials

This second example is of dynamic origin. I shall not discuss here the physical aspects of the phenomenon, but simply expose the following issue, which Leibniz explains very clearly (p. 662–664). In the considered problem, he and Huygens agree to consider that two of the essential physical quantities -namely time and space – are given by these two expressions:

$$ t = \int \frac{a^2 dv}{a^2 - v^2} \quad \text{and} \quad s = \int \frac{a \cdot v \cdot dv}{a^2 - v^2} $$

How should one compute these two quadratures? Leibniz then details to Huygens the differences between their two conceptions of calculus. For brevity, I will now analyze them by setting $a = 1$[8].

The first idea, commonly used by Huygens, is what we call 'series conception' (i.e., a calculus by summations of power series). The method comes directly from that of Mercator in his quadrature of the hyperbola, which Leibniz himself had used. Therefore, one has:

---

8   Obviously this is not a real limitation. Moreover, Leibniz himself later uses it in his letter (p. 663).

$$\frac{1}{1-v^2} = \sum_{n\geq 0}(v)^{2n} = 1+v^2+v^4+v^6+\ldots$$

Integrating term by term the "paraboloids", just like in squaring the circle (see above, Chapter II), one then obtains the value of t as the sum of a series (p. 662):

$$t-t_0 = \int\frac{dv}{1-v^2} = v+\frac{v^3}{3}+\frac{v^5}{5}+ \text{ etc.}^9$$

When Leibniz speaks about this equation to Huygens, he mentions it as "your expression". In doing so, he offers evidence that Huygens usually employs this method. As we know, it was also a favorite method of the English school. But Leibniz himself knew this practice, which he used abundantly in his *Arithmetical Quadrature of the Circle*.

On the other hand, the second expression that Leibniz proposes to calculate t is *directly* connected to his differential and integral calculus; this is a point he claims: this expression, he says, is "mine" (p. 662).

I describe briefly here a computation that Leibniz does not provide explicitly. For t, one uses a partial fraction expansion:

$$\frac{1}{1-v^2} = \frac{(1/2)}{1-v}+\frac{(1/2)}{1+v}$$

The summation of each term is separately performed:

$$\int\frac{dv}{1+v} = \log(1+v) \text{ et } \int\frac{dv}{1-v} = -\log(1-v)$$

$$t-t_0 = \frac{1}{2}\log\left(\frac{1+v}{1-v}\right) = \log\sqrt{\left(\frac{1+v}{1-v}\right)}^{10}$$

So, Leibniz obtains $\sqrt{\frac{1+v}{1-v}} = e^{t-t_0} = A\cdot e^t$ [11]

Thus, the *letteralized* exponential e^t is here introduced in a natural way. Moreover, this transcendent expression summarizes and encompasses the calculus by series. Leibniz insists to Huygens (p. 663, line 2) that both calculi (by series and differential) are equivalent: "You will find that both expressions match". Leibniz then argues in the same way with respect to s, which he first develops (first method) as a power series:

$$\frac{v}{1-v^2} = \sum_{n\geq 0}v\cdot v^{2n} = \sum_{n\geq 0}v^{2n+1} = v+v^3+v^5+v^7+ \text{ etc.}$$

9   t_0 is a constant of integration. In this text, Leibniz does not introduce such constants.
10   This is a local result, valid if $-1 <v <1$. The general result, valid if $v \notin \{-1,+1\}$, is written

$$\sqrt{\left|\frac{1+v}{1-v}\right|} = A\cdot e^t$$

11   To be clear for modern readers I chose to write e^A where Leibniz uses b^A.

By a similar term-by-term summation, one has, therefore, according to the series procedure:

$$s - s_0 = \frac{v^2}{2} + \frac{v^4}{4} + \frac{v^6}{6} + \text{ etc.}$$

But the second method (integral calculus) is here even simpler than for t! For Leibniz, it is indeed clear that $\dfrac{vdv}{1-v^2}$ is *directly, the differential of an expression*. One in effect immediately has:

$$s - s_0 = -\frac{1}{2}\log(1-v^2) = \log\frac{1}{\sqrt{1-v^2}}$$

$$\text{and so } \frac{1}{\sqrt{1-v^2}} = e^{s-s_0} = B \cdot e^s$$

This expression includes another *letteralized* exponential, which is thus equivalent to the power series above and which again summarizes it. These transcendent functions are, therefore, as natural as they are impossible to circumvent. Leibniz then extensively argues (p. 664–665):

> So, I have these exponential expressions, (that you call laughing *super-transcendent*) $\sqrt{1-v^2}$ for e^s and $\dfrac{1-v}{1+v}$ for e^t (…) I cannot understand why you find any obscurity in these expressions, because there is no more obscurity than in ordinary logarithms, which do not give you any trouble (…) I therefore believe that for curves going beyond ordinary algebra Equations, all you could wish in Analysis, is to express them by these new equations. Should we always do it, we would then perfectly know the nature of the curve, we could give its tangents, its quadratures, extensions, centers and even its intersections with a given curve, and solve by this way given transcendent problems; finally, I cannot see anything that should remain to be done after that, and all that would only require construction of logarithms, supplemented by constructions of ordinary geometry (…) If all these reasons are worth nothing, I was quite wrong in my calculus. I believed I transmitted you something very good, and of a great use.

In this letter dated 1690, addressed to Huygens – whose opinion matters so much to him – Leibniz was thus somewhat heavy-handed. His position, without nuances and which I have already analyzed above in Sections IV-C and V-D, is grounded, on the one hand, on the idealization in the calculus of *letteralized* exponentials and, on the other hand, on the legitimacy of the construction, when possible, of transcendent curves by ordinary geometry (the '*principle of percurrence*', in Leibniz's sense; see V-E).

Huygens' conversion

Leibniz's argumentation, supported by this second example, must have been especially convincing, as testified by Huygens' near-immediate reply of December 19th, 1690 (AIII, 4, N. 296, 682–691). This letter marks a clear inflection both on form and on substance. Huygens recognizes (p. 682), with a new and a somewhat unexpected humility, his lack of familiarity with Leibniz's new methods:

> You suppose that I fully understand all your calculus of Exponential and Logarithmic Equations; it is not the case; and so you organize a trial (...) before a judge who does not completely understand your language.

Shortly afterwards, the period of Huygens' skepticism finally comes to an end, concerning the vicissitudes of some curve (the catenary), and also as regards Leibniz's new inventions about the inverse method of tangents[12]. I cannot, here, analyze this in detail, but simply recall that, as soon as September 1st, 1691 (AIII, 5A, N. 36, p. 160), Huygens wrote to Leibniz:

> I then considered why I missed several of your discoveries, and I concluded that it might be a consequence of your new way of computing, which offers you, as it seems, some truths that you have not even sought.

A little later, September 17th, 1693 (AIII, 5B, N. 185, 634–635), one finds this acknowledgment, in due form, of the value of Leibniz's concepts and methods:

> I admire ever more the beauty of geometry in the new advances it makes every day, advances in which your contribution is so large, Monsieur, should it only be through your marvelous calculus. I am now a little bit acquainted with it, although I still do not understand anything of ddx, and I would like to know if you have encountered important problems where they should be used, so that I get desire for studying them.

---

12   On this point, see Gerhardt's preface to the Huygens-Leibniz correspondence. GM II, 3–10.

THE TRANSCENDENCE, BONE OF CONTENTION BETWEEN HUYGENS AND LEIBNIZ (1690)

XII-B A MATHEMATICAL STUDY

XII-B1 Huygens' geometrical problem

It is a problem of finding a plane geometrical locus, whose origin is not differential but classical (i.e., 'metric'). Huygens will explain the geometrical data to Leibniz in the fourth letter only[1], and not in the first one[2] which introduced the research. This will be the source of certain incomprehension on Leibniz's part. Here is the question in modern terms:

> $(A; x, y)$ is an orthonormal pair of axes. E is a fixed point on Ax (let a be its abscissa). Consider a variable line D with origin A. Let F be the point of D whose projection on Ax is E. To determine the locus of all points C of the plane such that $AC^2 = AE \cdot EF$.

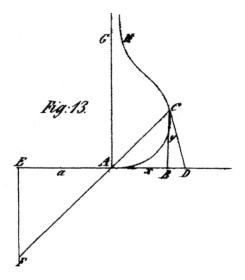

Figure 1

1   AIII, 4, N. 291, p. 654–658.
2   AIII, 4, N. 271, p. 545–550.

XII-B2 The Cartesian equation of Huygens' curve (H)

Let $Y = \lambda X$ the equation of D (we choose $\lambda \geq 0$), and x the abscissa of C. One has:
$C(x, y = \lambda x)$; $E(a, 0)$; $F(a, \lambda a)$.

One must have $AC^2 = AE \cdot EF$.

Therefore, $AC^2 = x^2 + y^2 = x^2 + \lambda^2 x^2 = x^2(1 + \lambda^2)$.

On the other hand, $AE \cdot EF = |a| \cdot |\lambda a| = \lambda a^2 = AC^2$.

Thus, $x^2 = \dfrac{\lambda a^2}{1 + \lambda^2}$. With $\lambda = \dfrac{y}{x}$, one gets for $x \neq 0$:

$$x^2 = \frac{\dfrac{y}{x} a^2}{1 + \dfrac{y^2}{x^2}} = \frac{\dfrac{y}{x} a^2}{x^2 + y^2} x^2 = \frac{x y a^2}{x^2 + y^2}$$

By dividing by $x \neq 0$, there remains:

$$x(x^2 + y^2) = a^2 y$$

(Huygens' curve (H) – relation (1))

Let us return on the excluded case $x = 0$. Then, $\lambda a^2 = 0$ and, therefore $\lambda = 0 = y$. It is thus the point A, and this point is also found in equation (1), which is the equation of a cubic curve. It admits A for a center of symmetry. To construct it, it is easier to determine a polar equation (with pole A).

XII-B3 The equation of Huygens' curve (H)
in polar coordinates

By setting $x = \rho \cos \theta$ and $y = \rho \sin \theta$, equation $x(x^2 + y^2) = a^2 y$ becomes: $\rho^2 = a^2 \tan \theta$. The curve is completely obtained for θ varying within any interval $[a, a + 2\pi]$. First, with $\theta \to \theta + \pi$, one finds the symmetry with respect to A. One thus constructs: $\rho = a\sqrt{\tan \theta}$ for $\theta \in \left[0, \dfrac{\pi}{2}\right[$

ρ increases from 0 to $+[\infty$ and $\rho(0) = 0$. There is a horizontal tangent in the pole, which is thus an inflection point. On the one hand, there exists an infinite branch $(\theta \to \pi/2^-)$. Is there an asymptote? Let us compute:

$$\lim_{\theta \to \frac{\pi^-}{2}} \rho \sin\left(\theta - \frac{\pi}{2}\right) = \lim_{\theta \to \frac{\pi^-}{2}} \left(-a\sqrt{\frac{\sin\theta}{\cos\theta}} \cos\theta\right)$$

$$= \lim_{\theta \to \frac{\pi^-}{2}} \left(-a\sqrt{\sin\theta \cos\theta}\right) = 0$$

The axis Oy is thus the asymptote. On the other hand, the intersection of (H) with the first bisectrix is a point Z with a vertical tangent with coordinates:

$$\rho = a \text{ and } \theta = \frac{\pi}{4}$$

At ths point Z, indeed, one has: $(\tan V)\left(\frac{\pi}{4}\right) = \frac{\rho}{\rho'}\left(\frac{\pi}{4}\right)$.

$$\rho' = \frac{a}{2\sqrt{\tan\theta}}(1 + \tan^2\theta)$$

Therefore, $\rho'\left(\frac{\pi}{4}\right) = a$

$$V\left(\frac{\pi}{4}\right) = \frac{\pi}{4} \text{ and } \alpha = \theta + V = \frac{\pi}{2}$$

At this point, $x = \frac{a}{\sqrt{2}} = y$, from which follows the form of (H):

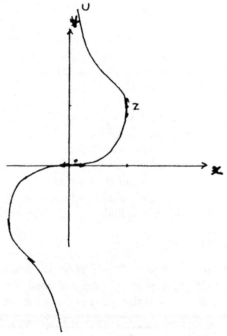

Figure 2

XII-B4 On calculations of the sub-tangent

Let us recall at first the modern definition and calculation of the sub-tangent to a curve (C) in one of its points M, with regard to an axis (which will here be Ox). The tangent at M intersects the axis Ox at T (see figure 3, below).

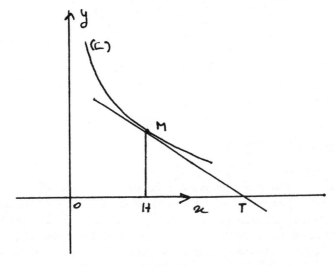

Figure 3

Under these conditions, the value of the sub-tangent is *conventionally* $\overline{HT}$. It is an algebraic value, that is, a real number of any sign[3]. This modern convention was also the one adopted by Huygens. Alternatively, Leibniz will take for its definition the opposite number, namely $\overline{TH}$[4]. This will be a source of the quarrel.

In modern terms, if we suppose that the curve is ***explicitly*** defined by $x \to y(x)$ (where y is a differentiable function), we thus have (with $Y-y = y'(X-x)$ and $Y = 0$ for the general calculation of the sub-tangent $\overline{HT}$):

$$\overline{HT} = X - x = -\frac{y}{y'} = -y\frac{dx}{dy} \quad \text{(Relation (1'))}$$

Note that, in the half-plane $y > 0$, the sub-tangent is thus negative if y is increasing and positive if y is decreasing.

---

3   Leibniz was well aware that the sign of the sub-tangent might fluctuate (cf. letter to Huygens dated 10 (20) April 1691, GM II, 90).

4   Neither Huygens, nor Leibniz uses the algebraic value notation. However, where Huygens notes BD for the sub-tangent (GM II, 46), Leibniz (GM II, 53) notes DB (?).

The "normal vector" method[5]

The above formula applies spontaneously in the case of a curve defined explicitly. But it can also be simply used in the case of a curve defined implicitly by F $(x, y) =$ 0, when F is a function of differentiability class C^1. We indeed have:

$$dF = 0 = \frac{\partial F}{\partial x}(x,y)dx + \frac{\partial F}{\partial y}(x,y)dy$$

$$\text{that is: } \frac{dx}{dy} = -\frac{\frac{\partial F}{\partial y}}{\frac{\partial F}{\partial x}}(x,y)$$

Therefore, as to the sub-tangent: $\rho = \overline{HT} = y\dfrac{\frac{\partial F}{\partial y}}{\frac{\partial F}{\partial x}}(x,y)$

This formula will be used in substance in various forms by many contemporaries of Leibniz, both for algebraic curves (F is then polynomial) as well as for certain transcendent curves whose implicit equations are known. It was therefore a fundamental method – we call it, in the following, the 'normal-vector' procedure –[6] for calculating the sub-tangent and, hence, to place the tangent in a large number of known curves at the time, both Cartesian (algebraic) and Leibnizian (transcendent).

What type of demonstration of the above result do we find among the contemporaries? They obviously did not use the notation of partial derivatives. The proof then involved classical geometrical issues (excluding any differentiation) for pure mathematicians, such as Huygens (see below), or alternatively it involved infinitesimal considerations that allowed the differentiation of F, and thus the calculation of the ratio $\dfrac{dx}{dy}$. This was the case with Leibniz, but also with Tschirnhaus and l'Hôpital.

The sub-tangent to Huygens' cubic curve

Huygens' cubic curve has equation $x(x^2+y^2) = a^2y$ (1). We have, by differentiating:

$$dx\,(3x^2+y^2) + dy\,(2xy-a^2) = 0, \text{ or else:}$$

$$\frac{dx}{dy} = \frac{a^2 - 2xy}{3x^2 + y^2}$$

---

5   Cf. *infra*, chapter XV, the section *Curves implicitly defined.*
6   This terminology refers to the fact that the tangent at a point M to an implicitly defined curve
     (by F = 0) is equally determined by the normal (i.e., the line orthogonal to the tangent and
     containing the point M); $\dfrac{\partial F}{\partial x}$ and $\dfrac{\partial F}{\partial y}$ are the coordinates of a normal vector at the relevant
     point (cf. Chapter VII).

Hence, relation (2): $\sigma = \dfrac{y(2xy - a^2)}{3x^2 + y^2} = \dfrac{2xy^2 - a^2 y}{3x^2 + y^2}$

In (2), we replace x^2 (in the denominator) by a value from (1):

$$x^2 + y^2 = a^2 \frac{y}{x} \qquad x^2 = a^2 \frac{y}{x} - y^2$$

Therefore:

$$\overline{HT} = \frac{2xy^2 - a^2 y}{3\left(a^2 \dfrac{y}{x} - y^2\right) + y^2} = \frac{2x^2 y^2 - a^2 xy}{3a^2 y - 2xy^2} = \frac{2x^2 y - a^2 x}{3a^2 - 2xy} \quad \text{(Relation (3))}$$

This expression (3) is, therefore, the one Huygens had submitted in the first letter to Leibniz – for a test – as the value of the sub-tangent[7]. He thus provided the value of the algebraic sub-tangent in the modern sense, the one given by relation (1') above. The value of Huygens' $\overline{HT}$ is not always positive: it is negative in fact on section AZ and positive on ZU. In conclusion, regardless of the method that Huygens used to calculate the sub-tangent (see below 7° section), it is as a number (not as a length) that he communicated it to Leibniz.

To be brief for the benefit of the modern reader, we preferred to use the differentiation of the equation of the curve for this calculation. Huygens, who refused the differential calculus, however, could not use this procedure. For him, this aspect was even an issue of the controversy with Leibniz! To obtain the sub-tangent, he had certainly used the 'classical' geometrical-algebraic method he exposed to De Witt in 1663; we describe it below at the end of this section.

So, Huygens finally asks Leibniz to find the equation of the curve (that he himself knows), of which the sub-tangent is equal to that obtained by relation (3), namely $s_0(x,y) = \dfrac{2x^2 y - a^2 x}{3a^2 - 2xy}$.

XII-B5 Leibniz's 'supertranscendent' curve

In the second letter[8], Leibniz answers by communicating his solution (p. 50) not to the geometrical original problem (which he still ignores at this moment) but to the one posed by Huygens, coextensive with the inverse method of tangents.

Far from the cubic (H) of Huygens, Leibniz finds the transcendent curve (L) with equation $\dfrac{x^3 y}{h} = e^{\frac{2xy}{a^2}}$ (relation (4)).

7   Huygens' notation is BD.
8   GM II, 49–52.

In the following letter[9], Leibniz proposes to justify this result by calculating the sub-tangent to (L) (p. 53). This method is usual to him, and it is interesting to follow his argumentation. On the relation (4), he first takes the logarithms:

$$3 \log x + \log y - \log h = 2x.y/a^2.$$

By differentiating this equation, he gets: $3\dfrac{dx}{x} + \dfrac{dy}{y} = \dfrac{2}{a^2}(x \cdot dy + y \cdot dx)$. By grouping:

$$dx\left(\frac{3}{x} - \frac{2y}{a^2}\right) = dy\left(\frac{2x}{a^2} - \frac{1}{y}\right).$$

And:

$$dx\left(\frac{3a^2 - 2xy}{xa^2}\right) = dy\left(\frac{2xy - a^2}{ya^2}\right).$$

Finally, the calculation of Leibniz gives: $\dfrac{dx}{dy} = \dfrac{2x^2y - a^2x}{3a^2y - 2xy^2}$.

To calculate the sub-tangent, however, he uses the formula $\sigma = + y\,(dx/dy)$, which is the opposite to that of Huygens. He thus gets:

$$+y\frac{dx}{dy} = \left(\frac{2x^2y - a^2x}{3a^2 - 2xy}\right) = s_0(x, y)$$

Thus, Leibniz justifies his transcendent solution since it admits the sub-tangent prescribed by Huygens!

XII-B6 The controversy (fourth and fifth letters)

In the fourth letter of the exchange (pp. 55–56), Huygens – finally – explains to Leibniz the question of geometry which was the source of the problem, and then communicates to him the result (1), which he found for the equation of the curve (H). He also gives him – still without any calculations, though the result is correct – the value of the sub-tangent of (H) obtained from these data (i. e., $s_0(x,y)$). There is thus an apparent contradiction, since both curves (H) and (L) – one algebraic and the other transcendent – appear to have the same sub-tangent. The misunderstanding between both protagonists comes entirely from the choice of the formula giving the sub-tangent. This formula, remember, is the result of a convention.

In the fifth letter[10], Leibniz rightly wants to clarify and resolve the paradox. He then gets to grips with the entire differential problem and, therefore, both curves (H) and (L), of which he calculates the sub-tangents: but he calculates, of course, with his own formula (pp. 62–63).

9   GM II, 53–55.
10   GM II, 59–64.

As such, he naturally obtains a couple of inverse results from Huygens: for Leibniz, Huygens' curve has for its sub-tangent $-s_0(x,y)$, while his own, the 'super-transcendent' one, would admit $+s_0(x,y)$ for its sub-tangent!

Leibniz insists, then: if one searches for a curve given by its sub-tangent, the sign of the latter is a major element – this point is indisputable, whatever the formula adopted for the sub-tangent[11]! However, and to show it better, Leibniz compares the two differential equations, which respectively define (L) and (H). This differential approach is quite natural to him. We give below both the differential equations concerned by adopting Leibniz's position as to the formula for the sub-tangent:

$$+y\frac{dx}{dy} = \frac{2x^2y - a^2x}{3a^2 - 2xy} \quad (E_1)$$

$$+y\frac{dx}{dy} = \frac{a^2x - 2x^2y}{3a^2 - 2xy} \quad (E_2)$$

From (E_1), one gets: $2xy^2.dx + 2x^2y.dy = a^2x.dy + 3a^2y.dx$.

Or else: $2(y \cdot dx + x \cdot dy) = a^2\frac{dy}{y} + 3a^2\frac{dx}{x}$. Actually, $2(ydx + xdy)$ is an exact differential, namely the differential of $2(x.y)$. (E_1) can therefore be integrated according to:

$$2(x \cdot y) = \log\left(y^{a^2}\right) + \log\left(x^{3a^2}\right) - \log K$$

In other words: $2x \cdot y = \log\left(\dfrac{\left(yx^3\right)^{a^2}}{K}\right)$. We thus find the relation (4), which provides

(L) – having defined $h = K^{\frac{1}{a^2}}$.

On the other hand, the development of (E_2), which involves $2(ydx-xdy)$, does not lead to an exact differential; the equation cannot be integrated by this method, as Leibniz points it (p. 63). One thus understands the reasons for the privilege granted by Leibniz to the differential equation (E_1) compared with (E_2) (that is, to (L) compared with (H), or else to the 'supertranscendent' curve compared with the cubic one).

---

11   The two differential equations $y\frac{dx}{dy} = \varphi(x,y)$ and $y\frac{dx}{dy} = -\phi(x,y)$ admit solutions which, *a priori*, can be radically different.

XII-B7 Huygens' method for the tangents
to algebraic curves

Huygens had detailed his method in a letter to de Witt dated February 25th, 1663[12], whose title was *Inventio Methodi ad Tangentes Linearum Curvarum*. He explained the determination of the sub-tangent to an algebraic curve (p. 312) on the example $x^3+y^3-axy = 0$ (that is, Descartes' *folium*): let B be the point of the curve with the abscissa x and ordinate y, and E be the point where the tangent at B intersects the axis. Huygens gives to x some increase (noted e). Let D be the point of the tangent in B at the curve of the abscissa $(x+e)$. By an application of Thales' theorem, Huygens writes the ordinate D as $y+y(e/z)$, where, he says, we have z = FE. This is the value of the sub-tangent of the curve at B – however, *in absolute value*.

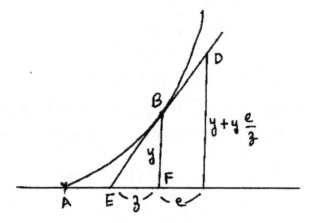

Huygens injects, then, both increases $x \rightarrow x+e$ and $y \rightarrow y + y\,(e/z)$ in the equation of the curve. The method, similar to that of Fermat, consists in keeping only the terms of lower degree with respect to e (i. e., whose degree is less than or equal to 1) in the obtained development, and in neglecting the higher powers. By equating to zero all such terms, we can calculate z as a function of e, and then, as a second step dividing by e, find the value of z[13]. Huygens gets (p. 314):

$$z = \frac{-3y^3 + axy}{3x^2 - ay}$$

This is the absolute value of the sub-tangent. By elementary geometrical considerations, Huygens then discusses the sign in order to find the algebraic value (the 'real

---

12 [Huygens 1888].Tome IV, 312–317.
13 $(x+e)^3+y^3(1+(e/z))^3$-a$(x+e)$y$(1+e/z) = 0$. By keeping only the terms whose degree is 0 or 1 (with respect to e):
$x^3+3x^2e+y^3(1+3z/e)$-ay$(x+e+ex/z) = 0 = x^3+y^3$-axy+e$[3x^2+(3y^3/z)$-ay-$(axy/z)] = A+B$.
A is 0. In B, there remains $z = -\dfrac{e(3y^3 - axy)}{e(3x^2 - ay)} = \dfrac{axy - 3y^3}{3x^2 - ay}$

value'); he thus changes the sign (p. 317), and finally finds for the sub-tangent the value:

$$\sigma = \frac{3y^3 - axy}{3x^2 - ay}$$

This result, which can easily be found by the standard method of the normal vector (see above), is correct. Without using differential calculus, Huygens' method thus enabled the construction of tangents to implicitly defined plane algebraic curves. Generalizing the example of the *folium*, the procedure is then described in a rule (*Regula*, pp. 315–317).

CHAPTER XIII
TRANSCENDENCE: THE WORD AND THE CONCEPTS, FROM EULER AND LAMBERT TO HILBERT

XIII-A EULER AND TRANSCENDENCE

Euler and the classification of functions:algebraic or transcendent

The two volumes of the *Introductio in Analysin Infinitorum* of Euler (1748)[1] are, in many respects, an updated and exhaustive review of the Leibnizian concepts of transcendence. The *Introductio* is important for this volume; however, I can only make a brief commentary here. In the preface (vol. I), Euler defines his strategy regarding the "functions of variables":

> Initially, I have divided these into algebraic and transcendental functions; the first of which are formed by the common operations of common algebraic from variable quantities, these truly either are composed by other accounts or are effected by the same operations repeated indefinitely[2].

Later in the preface Euler writes about several transcendent functions:

> The series are of such a kind that their sums are expressed by logarithms or by circular arcs. ([Euler 2013], vol. I, 2)

Dealing with algebraic functions (Chapter I, page 2), Euler then notes:

> The first subdivision of algebraic functions shall be into rational and irrational functions.

It should be noted that this division between rational and irrational algebraic functions, even if it is correct and apparently natural for the functions of one variable, raises some compatibility issues as to the nature of the functions of two variables (see *infra* chapter XV)[3].

Nevertheless, Euler's definition above is certainly one of the first clear classifications of the functions in mathematics[4]. Returning to the foreword of the *Introductio* and to the sentence quoted above, one may conclude concerning Euler's divi-

---

1 I mainly use the English translation [Euler 2013] by Ian Bruce of the original work [Euler 1748]. However, I noted everywhere "transcendent" where Bruce noted "transcending".
2 [Euler 2013], vol. I, 2.
3 For example, y: $x \rightarrow \sqrt{x^2 - x} = y$ is certainly an irrational algebraic function of one variable, according to Euler. However, one has $0 = F(x,y) = x^2 - y^2 - x$ – and F is this time a function of two variables (a polynomial); it is thus algebraically rational (for Euler). Therefore it might be better to use a more satisfactory definition of rational algebraic functions of one variable, a point that will be analyzed in more detail in Chapter XV.
4 On the history of the genesis of the concept of function, see [Youschevitch 1976]. Here, pages 61–62.

sions of functions. On the one hand, there are algebraic functions, rational or irrational; on the other hand, there are non-algebraic functions, which Euler calls "transcendent".

Note, incidentally, that the division of functions between algebraic and transcendent is not the only one in the *Introductio*. In Chapter I, other categorisations that are often modern and full of interest come to overlap: explicit functions / implicit functions (for irrational algebraic functions, see vol. I, 10, § 8)[5] and uniform functions / multiform functions (pages 11–12).

Euler, algebraic curves and transcendent curves

In a natural way, this inventory of functions involves a correlative geometrical inventory, namely of curves, algebraic or transcendent, according to their equation. Note that Euler does not consider curves devoid of equations (unlike Leibniz). This classification of curves is the subject of a comprehensive study – not, however, in Book I but in Book II, which is dedicated to geometry[6]. For the first time in history, a specific work was thus dedicated to the transcendence of functions and curves: Chapter 21, "Concerning transcendent curved lines"[7].

By 1748 the use of both the terminology and the Leibnizian conceptualization of transcendence of curves and functions was no longer problematic for Euler. This daily and easy use of these in the text thus gave mathematical right of citizenship to Leibnizian transcendence. Euler made considerable use of Leibniz's terminology on transcendence; he never references the "geometrically irrational" curves of Newton. It is thus Euler who introduced and extensively spread Leibnizian transcendence in the mathematical community, from his own period in history until the present moment.

In the first chapter of vol. II[8], Euler exposes first, as clearly as for the functions, his conception of the division between algebraic and transcendent curves (see quotation below)[9]. In Euler's words:

> Therefore since a knowledge of curved lines leads to functions, we may consider that there are just as many kinds of curved lines present as there are functions above. Therefore curved lines may be divided most conveniently into *algebraic* and *transcendent* according to the manner of functions. Clearly a curved line will be algebraic, if the applied line y were an algebraic func-

---

5   The distinction is interesting. Euler explains: "The functions are *explicit,* which have been set out through root signs, examples of this kind have been given just now. The functions are *implicit,* which truly are irrational functions, which have arisen from the resolution of equations" (page 10). If he had persevered using this definition, Euler could have given a satisfactory definition, 'in Lambert's style', of transcendent numbers.
6   See [Breger 1986], 132, n. 88.
7   [Euler 2013], vol. II, 564–604.
8   This passage is quoted by Breger ([Breger 1986],132, n. 88).
9   From this point on, Euler now uses continuously the 'functional' terminology' that he introduced in book I.

tion of the abscissa x^{10}; or, since the nature of the curved line may be expressed by an algebraic equation between the coordinates x and y, any curved lines of this kind are accustomed to be called also *geometric*. But a curved line is *transcendent*, the nature of which is expressed by a *transcendent* equation between x and y, or from which y is a *transcendent* function of x. And this is the particular division of continued curved lines, since these are either *algebraic* or *transcendent*[11].

This excellent definition of the transcendence of curves finally reaches a precise point, after the many uncertainties of the previous century, as described above. This chapter is epoch-making in the history of the transcendence of curves and functions.

An extensive overview of transcendent functions

Much of the chapter is devoted to transcendent curves (and functions). Euler first gives some basic examples of transcendent functions (vol. II, p. 564, § 507):

$$x \to \log x;\ x \to \sin x;\ x \to \cos x;\ x \to \tan x$$

then (page 565; § 509): $x \to x^{\sqrt{2}}$. Euler, in a relevant reference to Leibniz, notes that the latter had called *interscendent* this exponential (see IV-C3, "Towards a hierarchy in transcendence: the interscendent exponentials"). Concerning this example, he even writes:

"Talis aequatio autem nulla via geometrica potest".

In these circumstances some curves defined by the only symbolism of their equations will have, for Euler, definitively given up the field of geometry (on this point, see some similar conclusions above)[12]).

Euler (pages 566–567, § 512) then attempts a classification according to the symbolic complexity between transcendent curves. He thus classifies "in the first row" those whose equation contains, besides the algebraic functions, only logarithms. It is a natural process of categorisation of transcendence already made, under diverse forms, by Leibniz and John Bernoulli.

Besides logarithms, Euler adds (pp. 572–574) the letteralized exponentials $y=x^x$ (p. 572; § 518). Accordingly, he shows an extensive interest in the curve (C) of equation $x^y=y^x$ (p. 573; § 519), a truly paradigmatic transcendent curve, via two exponentials, fully in line with the 1682 discussions between Leibniz and Tschirnhaus (cf. XI-A). The complete treatment of this curve is remarkable because Euler

---

10   We know today that it is not exactly the same issue: if $y=F(x)$, where F is algebraic (in the Eulerian sense), one then has $G(x, y)=0$ (where, in the same way, G is an algebraic function in the Eulerian sense). But we cannot assert the reverse: if y is such as G $(x, y)=0$, we cannot assert that y is algebraic (in the Eulerian sense). It may happen that some solutions (with respect to y) of $G(x,y)=O$ are not expressible by radicals.

11   [Euler 2013], vol. II, 5 (§ 15).

12   Chapter IX-D, "The advent of the transcendence and the detachment from geometry as the only guarantee of the truth".

clearly gives, for one of very first times in history, *parametric representations*[13]. This is fundamental to the history of mathematical ideas. Euler first notes that the curve of equation $x^y = y^x$ obviously contains, in particular, the half-line D of equation $x = y$ (and $x > 0$). But he then pertinently notes (p. 573; § 518) that, "the proposed equation has a wider meaning". By putting $y = tx$, one actually has:

$$x = t^{\frac{1}{t-1}} \quad \text{and} \quad y = t^{\frac{t}{t-1}} \quad (1)$$

However, Euler notes, by putting $t - 1 = \frac{1}{u}$, we also have:

$$x = \left(1 + \frac{1}{u}\right)^{u} \quad \text{and} \quad y = \left(1 + \frac{1}{u}\right)^{u+1} \quad (2)$$

Either parametric representation (1) or (2) thus allows for the easy construction of transcendent curve (C). It is necessary, however, to add the half-line D.

In these pages Euler actually focuses – by following Leibniz – on the widest possible overview of transcendent curves known at the time. And this inventory, which examines all curves proposed by Leibniz, Bernoulli, Newton, etc., is remarkable. After the exponentials, Euler is interested in reciprocal trigonometric functions. First, the function $x \to y(x)$, as it is defined by $x = \sin y$ (page 574; § 520) (such a function is written today as $y(x) = \text{Arc sin } x$). He makes in this regard a reference to Leibniz (page 575). He then continues with $x = \tan y$ (page 576; § 521), and follows with $x = \dfrac{1}{\cos y}$.

He then looks at the cycloid (ibidem), defined by a Cartesian equation, and then epicycloids and hypocycloids (p. 578; § 523), which he defines in a common geometrical way but for which he provides, once again, *parametric* equations (p. 579), a practice which, I repeat, was at that time as remarkable as it was new. Further, he concentrates on a study of various spirals (the beginning is on page 583; § 526): the Archimedean spiral; the hyperbolic spiral of Bernoulli (p. 584; § 527); the lemniscate of the same Bernoulli (p. 584; § 527); and, finally, the logarithmic spiral (idem).

Euler, algebraic quantities and transcendent quantities

As we have seen, the classifications provided by Euler of functions as well as of curves were in every respect remarkable. Let us repeat: this chapter of Euler is epoch-making in the history of the transcendence of curves and functions, both for its precision as much as its methodological inspiration. What may be less accurate

---

13   Recall that this completely new conception, namely the symbolic description of a plane curve through two functions with regard to a single parameter, had hardly appeared spontaneously to Leibniz (this point is, however, controversial). See above, V-B), "Leibniz's critique of Descartes' geometry 'of the straight line' and the parameterized curves".

is doubtless the classification of *transcendent quantities*. Let us first revisit a part of the preface already quoted above:

> The series are of such a kind that their sums are expressed by logarithms or by circular arcs; which since they shall be transcendent quantities, are shown in turn by the quadrature of the hyperbola and of the circle.

For a modern mathematician, the interpretation of this essential sentence may raise a problem with the two occurrences of the word "transcendent" that do not seem directly related. At first, the expression "The series are of such a kind that their sums are expressed" can be understood either in a functional sense or in a numerical sense. Given the previous context, the functional sense is the most likely[14]. However, this assumption is made immediately problematic with the sentence that follows: these sums are, Euler says, *quantities* connected with the area of the circle or the hyperbola.

In the first part of the book there is not a precise definition of an algebraic quantity or a transcendent one. Euler seems to consider that these conceptions are taken for granted. One does not find them further in Chapter VI, "Concerning Exponential And Logarithmic Quantities"[15], which considers a^z without comment on its status as a transcendent quantity. And one does not find any further definitions in Chapter VIII, which is nevertheless entitled "On transcendent quantities arising from the circle".

From the opening of the book there appears a dialectical shift from "transcendent function" correctly specified to "transcendent quantity", which could be defined, implicitly, only by an analogy. In fact, a quantity that is written *via* a transcendent function in the Eulerian sense is not necessarily transcendent[16].

On quantities that are not expressible by radicals

Let me repeat: for Euler, the definition of algebraic and transcendent quantities seems to be obvious. In another work[17], I developed what seems to me to be Euler's implicit representation of a transcendent number, namely a quantity that would admit no expression written with the usual symbols (including the radicals). In short, "transcendent" means for Euler that which is "not expressible by radicals". For Euler, this implicitly supposes on the one hand that *any quantity can be written* (this relationship to the symbolism is constant for Euler) and, on the other hand, that if a quantity admits an expression by means of a transcendent function then it cannot admit another one that could be formed with the usual (i. e., Cartesian) symbols. It

---

14 One can also believe that Euler directly follows Leibniz's perspectives: e^x, for example, is a transcendent *function*.

15 [Euler 2013], vol. I, 152.

16 For more details on this point, the reader should consult [Serfati 1992], supplement N°2: *Émergence de la notion de nombre transcendant chez Euler*, 175–176.

17 See [Serfati 1992].

is thus a transcendent quantity. One finds here a thought pattern similar to Gregory's.

The rest of Euler's work outside of this specific publication contains clarifications and possible confirmation of this hypothesis. This interpretation is indeed validated by a late article, "De relatione inter ternas pluresque quantitates instituenda"[18], published in 1785 in the *Opuscula analytica*. In this text, Euler first devotes himself to show that π cannot be a quadratic irrational. The argument is simple: its continued fraction expansion is not periodic. Another supplementary issue fundamental for Euler's work on transcendence is: does π admit or not an expression with radicals (other than quadratic)?

In all cases, however, the central issue for Euler is, with regard to π, that of the possibility or impossibility of writing it with radicals, whether quadratic or not. In the words of Euler:

> However, whether transcendent quantities, such as those which involve the circumference of a circle or logarithms, are able to be compared even with any sort of algebraic quantity, seems at this point most uncertain, if indeed such an impossibility has been shown by no one. Indeed, it seems demonstrated so greatly enough that the perimeter of a circle with diameter 1 allows no comparison with simple quadratic radical formulas, since otherwise, a continued fraction, equal to π itself, ought to have periodic terms, a thing which does not seem to happen at all. We are compelled to leave in doubt (*in dubio relinquere cogimur*) whether the quantity π is able to be compared with formulas arranged such (...)[19].

Later in the same text (§ 11 and § 12), Euler tries vainly to find a linear relation with rational coefficients between v2, v3 and π, by following the approach that is precisely the object of his article ("De relatione inter ternas (...)"). He gives up after several laborious calculations, then calls naïvely upon an argument of simplicity: if there were an exact relation, this one could only be simple. Yet it is not so[20]! From the fact that π cannot be simply connected (and therefore written) *via* $\sqrt{2}$ and $\sqrt{3}$, he "infers" that it probably cannot be with any other radical and finally concludes that π belongs to a "special" kind of transcendent quantities. One can legitimately believe that this statement of principle by Euler (with regard to π) derives directly from the part that π vehicles in the issue of squaring the circle. π will remain, in this sense, a crucial issue in the transcendence of numbers.

---

18   [Euler 1785]. I mainly use the English translation [Euler 2010] by Geoffrey Smith, *On Establishing a Relationship Among Three or More Quantities*. However, I noted everywhere "transcendent" where Smith noted "transcendental".

19   [Euler 2010], 6–7, § 10.

20   This argument may today seem very naïve; it was nevertheless common at the time. Lambert will also use the same naive dialectics of "proof by simplicity". See [Lambert 1761]. Cf. *infra* XIII-B).

A hierarchy of transcendent numbers?

The end of the text ([Euler 1785], 142, § 12) = [Euler 2010], 8) suggests – in an intuitive and implicit mode – some form of hierarchy of transcendent quantities, since the same π being supposed transcendent by Euler is also supposed not to be composed of other transcendent quantities that would beforehand have been defined. In the words of Euler:

> I will not continue these operations further, since if an exact relation were to be given, without doubt it would not be so complicated. Moreover it would have been of little use to provide such nearly true relations. From this point, the idea seems to be certain enough, because the perimeter of a circle should establish such a type of transcendent quantities that it may allow itself in no way to be compared with any other quantities, whether surds ([21]) or transcendents of another type ([22]).

Transcendent numbers have a very extensive domain

Euler ([Euler 1785], 143, § 13) reaffirms the conception of a progressive complexity in transcendent numbers, as well as their (considerable) extent:

> Moreover, there exist other infinite classes of transcendents, which are not able to be reduced either to a circle nor to logarithms, even if they should seem to hold any relation with those numbers. If by chance such quantities were to have any exact relation with some hitherto known number[23], which one may not define directly from analytic principles, this method[24] seems to supply a unique way by which one, with a benefit just as intuition, may investigate relations of this sort ([Euler 2010], 8).

Regarding the extent of transcendent numbers, Euler is here quite prophetic. He is indeed far from considering, as one might spontaneously imagine, that such numbers are exceptional singularities, which would be more or less directly associated with logarithms or with circular functions, as e or π. On the contrary, he describes them as an "infinity of classes". In this, Euler announces Cantor and his proof of the uncountable character of the set of transcendent numbers, unlike the rational. In the following paragraph, ([Euler 1785], 143), he gives the example of $a = \sum_{n\geq 1} \dfrac{1}{n^3}$.

> § 14...Therefore, I will fairly accurately present here a single example of this sort for which such a relationship might exist. Consider the sum of a reciprocal series of cubes (…) which I was hitherto able by no means to reduce, whether to a circle or to logarithms, although never-

---

21   The term "surds" meant at the time what we now call algebraic irrationals, the expression of which contained mostly radicals.

22   *Cum nullis aliis quantitatibus, sive surdis, sive alius generis transcendentibus, nullo modo se comparari patiatur* page 142.

23   "*Quandum translated as quandam*". Note from the translator.

24   The "method" mentioned here is that of Euler in this article, as explained above: search naively, blindly ('almost by guessing it'), a linear combination involving the transcendent quantities concerned, with several other quantities (algebraic irrational or transcendent) already known.

theless the sum of all the second powers is able to be produced through second powers of π itself, and moreover the alternating sum of the first powers

$$1 - \frac{1}{2} + \frac{1}{3} - \frac{1}{4} + \frac{1}{5} - \text{ etc.}$$

expresses the logarithm of 2^{25}.

The value a of this sum above is thus another transcendent number of a new kind ([26])! The issue of the inventory of the transcendence that so occupied Leibniz – rightly so because he was the inventor – is here downgraded and sent back to the unknown domain of "infinite classes".

Euler: proofs of irrationality and conjectures of transcendence

The situation can be thus summarised at Euler in the second half of the eighteenth century. He sprinkles his texts, firstly, with demonstrations of irrationality – genuine or naïve – and, secondly, with conjectures of transcendence, often implicit and in varying degrees. In the conjectures concerned, the term "transcendent" must obviously be taken in the "Eulerian" sense. Such is, for example, Euler's famous "conjecture of transcendence", which I explain below.

Regarding irrationality, Euler could obviously use the continued fractions, which were at the time the Alpha and Omega of any proof of irrationality. In terms of continued fractions, irrationality, which is a result of the "failure" of the process of the Euclidean division, is easily inferred from the non-finiteness of the expansion. Thus, since the continued fraction expansions of π and e are not zero from some rank, Euler obviously concludes their irrationality. Moreover, because these expansions of both numbers are not even periodic, he can infer that they are not more quadratic irrational. Such conclusions, however, would not exclude the fact that π could be a more complicated algebraic number. But, as we saw above, Euler is probably persuaded that π is of another nature, namely that it is truly a *transcendent number.*

In Chapter VI of the *Introductio*[27], Euler proposes another conjecture of transcendence, which went on to become famous. It is a question this time of logarithms and can, in modern terms, be formulated as follows: if a and b are two rational numbers, possibly two algebraic numbers, then either $\log_a b$ is rational (this is the case when, Euler notes, b is a "power of the base") or it is not even algebraically irrational and is thus transcendent. For instance, $\log_2 3$ is transcendent and

$\log_{2\sqrt{2}} 32$ is rational. Because $\log_{2\sqrt{2}} 32 = \dfrac{\ln 32}{\ln 2\sqrt{2}} = \dfrac{\ln 2^5}{\ln 2^{\frac{3}{2}}} = \dfrac{5\ln 2}{\frac{3}{2}\ln 2} = \dfrac{10}{3}$. This con-

---

25 [Euler 2010], 8. This result was obtained by Mercator.
26 On the following page, Euler uses this method to look for an approached linear relation between a above and two already known sums of other numerical series.
27 *Concerning Exponential And Logarithmic Quantities* = [Euler 2013], vol. I, 152.

jecture, which much later (1900) became the subject of Hilbert's "Seventh Problem" was completely solved by Gel'fand (1929 and 1934)[28].

XIII-B LAMBERT AND TRANSCENDENT NUMBERS

Just like Euler, Lambert was interested in transcendence, however from different perspectives and with widely different centres of interest. Accordingly, he did not attempt to give, as did Euler, a general classification of functions and curves. He dealt, on the other hand (very pertinently), with numerical transcendence. This study was the conclusion to his famous *Mémoire* (Report) where he demonstrated, for the first time in history, the irrationality of π ([Lambert 1761]). Below, I will highlight two points:

–   On the one hand, the *Mémoire* in question is based on certain results directly provided by Leibniz's work (see "Introduction" below); and
–   Secondly, the classification of numbers given by Lambert in *Mémoire* is the source of *all modern theory of transcendent numbers*, from Hermite to Lindemann and Siegel (see "Conclusion" below).

Lambert's method for proving the irrationality of π is quite remarkable. In another work ([Serfati 1992]), I epistemologically analyze the method in detail; here, I summarize some of my conclusions. Lambert first considers the power series expansions, common at the time, of sin V and cos V, where V is a rational number. He infers, by quotient, the *infinite continued fraction expansion* of tan V[29]:

$$\tan V = \cfrac{1}{W - \cfrac{1}{3W - \cfrac{1}{5W - \text{etc.}}}} \quad \text{where} \ \ W = \frac{1}{V}$$

Lambert demonstrates *rigorously* by induction (p. 116–118) the values of the coefficients of this continued fraction; these are simple in contrast to those of the power series of tan V. Lambert then reveals his mathematical strategy:

> The issue is to show that every time an arc of a circle is commensurable with the radius, the tangent of this arc is incommensurable with it, and conversely any commensurable tangent is not the one of a commensurable arc. (p. 113)

In other words, V and tan V cannot both be rational. Lambert continues, unexpectedly:

> This statement seemed likely to admit an infinite number of exceptions, and it admits none.

---

28   On the future of this conjecture, see section XIII-D) in the present chapter. I also refer the reader to [Serfati 1992], both to paragraph 20 (*'David Hilbert'*), pages 163-166, and to the supplement, (*'Émergence de la notion de transcendant chez Euler'*), pages 175-176.
29   This result is thus functional (V is a variable), not just arithmetical. We will call it a *generalised continued fraction*.

Lambert does not add here (he does so later on) how he will very simply use his result: as tan $(\pi/4)=1$ is rational, then $\pi/4$ is not, nor π. I shall not comment further here on Lambert's proof; I will focus only on the introduction and conclusion of *Mémoire*.

Introduction to *Mémoire*: on "proof by simplicity"

Lambert wishes first to analyze the intuitive fact that the ratio of the circumference to the diameter (that is, π) cannot be rational. The text opens with these lines: "Demonstrating that the ratio of the diameter of the circle to its circumference is not as an integer is to an integer, this is a thing of which geometers will hardly be surprised". Then he continues with the following naïve "proof by simplicity":

> We know Ludolph's numbers, the ratios found by Archimedes, by Metius etc., and a large number of infinite sequences which relate to the squaring the circle. And if the sum of these sequences was a rational quantity, we rather naturally had to conclude that it will be an integer, or a very simple fraction. Because if it was a very composed fraction, why would it be this one rather than such other one?[30]

In short, for Lambert, if π is not a simple fraction, it cannot be a fraction. To be better understood, he immediately adds:

> Thus, for example, the sum of the series:

$$\frac{2}{1.3}+\frac{2}{3.5}+\frac{2}{5.7}+\frac{2}{7.9}+ \text{ etc.}$$

> is equal to the unit, which is the simplest of all the rational quantities. But if we alternatively delete 2, 4, 6, 8, etc., terms, the sum of the others

$$\frac{2}{1.3}+\frac{2}{5.7}+\frac{2}{9.11}+\frac{2}{13.15}+ \text{ etc.}$$

> gives the area of the circle the diameter of which is equal to 1. It thus seems that if this sum is rational, it should also be able to be expressed as a very simple fraction such as $3/4$ or $1/2$. Indeed, the diameter being equal to 1, the radius to $1/2$, the square of the radius equal to $1/4$, it is clear that as these terms are so simple they cannot be obstacles. And as it is with the circle, which is a kind of unity, it is clear that, in this respect, we cannot expect a very composed fraction ([Lambert 1761], 112–113).

Lambert does not state from which numerical series he found the results[31]. Yet these are unquestionably imported from Leibniz, one by "the sum of all differences"[32],

---

30 [Lambert 1761], 112.
31 The fact that both results involve a passage to a limit is evidenced by "+ &c".
32 As for the first result, one immediately acknowledges Leibniz's method of *the sum of all differences* $\left(\left(\int dx\right)\right)$ (cf. IX-B and [Serfati 2013b]). The value of the sum is indeed
$$\left(1-\frac{1}{3}\right)+\left(\frac{1}{3}-\frac{1}{5}\right)+\left(\frac{1}{5}-\frac{1}{7}\right)+ \text{ etc. } =1.$$

the other as a by-product of the arithmetical quadrature of the circle[33]. Let us repeat, the argument is of simplicity: if, from the sum of a series equal to the unit we re-move a term on two and if the remaining result ($\pi/4$) continues to be a rational number, then this one could be only very simple, in particular as for its denomina-tor. Lambert "does not see" where from the complexity may originate, from such simple data and from a result that is attached to the length of a simple figure – namely the circle, which is a "kind of unity"!

Beyond these intuitive and somewhat "psychological" considerations, Lambert then makes a beautiful and *rigorous* mathematical proof of the irrationality of π (after which he will move on to e) using generalised continued fractions, which I shall not study here (cf. [Serfati 1992], 62–83). He then establishes his central re-sult, which may be stated in more modern terms as follows: "V and tan V cannot together be two rational quantities". This allows Lambert to demonstrate the irra-tionality of π.

It is then easy for Lambert to replace circular functions with hyperbolic ones in the series expansions, that is to say, sin V by sh V and cos V by ch V. Where there was previously an alternation of sign, there remain positive signs only. By quotient, Lambert also easily achieves a generalised continuous fraction expansion of tanh V (and thus immediately of e^V) and the conclusion: "V and tanh V cannot together be two rational quantities". Accordingly, if V is rational, tanh V is not, nor e^V. By set-ting V=1, we thus see that e is not rational, nor e^2, etc. In the words of Lambert:

> This lets us know to what extent the irrationality of the number e is transcendent because none of its powers, nor any of its roots can be rational.

Concerning Lambert's fundamental result, he also pertinently adds:

> This statement shows to what extent the transcendent quantities are transcendent and remote beyond any commensurability. (p. 113)

In the two preceding quotations, the term "transcendent" has no precise technical meaning in Lambert; it simply means that the quantities involved are irrational in an extraordinary way, beyond any standard.

> I now turn to the last pages of *Mémoire*.

---

33   As for the second result, the partial fractions expansion:

$$\frac{2}{x(x+2)} = \frac{1}{x} - \frac{1}{x+2}$$

provides this expression of the sum:

$$\frac{1}{1} - \frac{1}{3} + \frac{1}{5} - \frac{1}{7} + \frac{1}{9} - \frac{1}{11} + \text{ etc.}$$

and this is (again) the Leibnizian series obtained in his quadrature of the circle (called today the *Arctan serie*). The value of the sum is $\pi/4$.

Conclusions of *Mémoire*: a classification of numbers

Mémoire ends with two brief and important pages where Lambert first exposes his own distinction between quantities, rational, irrational and transcendent (in the modern sense). We saw in the previous section that Euler had also worked on numerical transcendence, even proposing conjectures of transcendence. However, even if he frequently uses the term "transcendent quantities", he does not define them with sufficient precision. Lambert will be much more explicit here.

In the very last pages of Lambert's text (§ 89, 158), he first reveals his own distinction between quantities, supported by the two examples just outlined (i. e., *e* and π). For Lambert, there are at first the quantities that are "commensurable with the unit" (that is to say, the rationals) and among those that are not (the irrationals), he formulates a new categorisation: on the one hand, there are the irrational quantities that are written with radicals and also those that are irrational roots of algebraic equations. Lambert places these two types of quantities in the same category (this point is epistemologically decisive) and he calls them both ***radical irrational quantities***.

On the other hand, and always among the incommensurable quantities, there are obviously still the others (by which, however, Lambert started his distinction!), namely those that are not even what he called "radical irrationals". They are quantities, he says, that are formed "at random" (*"au hazard"*) and about which we do not know much, not even the means to study them. They are hardly "within the field of the analysis". Here is Lambert (§ 89, page 158, emphasis in the original):

> All that I have just exposed on the circular and logarithmic transcendent quantities appears to be based on much more universal principles, but which are not yet developed enough. Here is however what can be used to give some idea. It is not enough to have found that these transcendent quantities are irrational, that is incommensurable with the unit. This property is not unique to them. Because, besides that there are quantities which we can form at random, and thus are hardly within the field of the analysis, there are still an infinity of others we call *algebraic*: such are all the *radical irrational quantities* such as
>
> $$\sqrt{2}, \sqrt{3}, \sqrt{4}, \text{ etc. } \sqrt{\left(2-\sqrt{3}\right)} \text{ etc., and all the } \textit{irrational roots of algebraic equations, } \text{such}$$
>
> as those of the equations
>
> $$0 = xx - 4x + 1$$
> $$0 = x^3 - 5x + 1$$
> $$\text{etc.}$$

The ones and the others, I will call *radical irrational quantities*.

We shall thus note that for Lambert, a "radical irrational quantity" is what we now call an algebraic number (while it is not necessarily expressed with radicals!). We should also note that what Lambert calls a number "at random" is what we call a transcendent number today. If the concepts are therefore well distinguished, the terminology is not that which is retained today. However, Lambert will partially return to this point.

Here we discover a discriminative and creative statement that distinguishes between two categories in the irrationals, starting from an effective and theoretically simple criterion: it is not any more necessary to envisage (as Euler) all the conceivable radicals but "only" all equations with integer coefficients, which is obviously simpler. Lambert here seems to be the first to give this precise definition of numerical transcendence. Of his predecessors, the figure most likely to have approached this was probably James Gregory in *Vera Circuli … Quadratura* (see above II-B).

This categorisation is epistemologically fundamental. It shows the supremacy of Lambert's conception over that of Euler. At the same time it displays the conception of the present modern theory of transcendent numbers. Lambert then pursues his conclusion:

> I say that no circular or logarithmic transcendent quantity can be expressed by some radical irrational quantity that is related to the same unit and in which is not present a transcendent quantity. (§ 90)[34]

This result, which Lambert thinks "can be demonstrated", can thus be interpreted today as follows: the numbers e and π, which are non-algebraic in the Lambertian sense of course, are also non-algebraic in the Eulerian sense – that is to say, not expressible by radicals.

In the last lines of his text, Lambert then focuses on issues of constructibility with ruler and compass. He explains that he understands that all which is constructible is algebraic but that all which is algebraic is not necessarily constructible[35]:

> It happens only very rarely that the latter [the radical irrational quantities] […] can be arbitrarily constructed. (page 159, § 91)

Finally, Lambert postulates in an almost prophetic way the impossibility of squaring the circle *via* the transcendence of π:

> Since the circumference of a circle cannot be expressed by some radical quantity nor by some rational one, it follows that there will be no way to determine it by some geometrical construction. Because all that we can construct geometrically involves rational and radical quantities.

The procedure described here is precisely the one that Wantzel, then Hermite and then Lindemann will offer (see XIII-D) below. Lambert appears here as an irreproachable visionary.

On the supremacy of symbolism and issues of 'transcendent' terminology

The central point of the supremacy of Lambert's definition is symbolic: we can write symbolically and easily *any* equation with integer coefficients but not *any* quantity written with sums, differences, products, quotients and radicals (even if these quantities are strictly included in the set of the solutions of those equations). Here, one discovers two different symbolical positions of the mathematician. That Lambert's definition surpasses that of Euler and only permanently survived thus reflects the supremacy of symbolism.

I also pointed out above that in his first founding classification, Lambert uses the term "number at random" to refer to what we now call a "transcendent number".

---

34   In this passage the term "transcendent" is to be taken in the simple sense of "outside the standard".
35   See our study, *Wantzel, la règle et le compas*, in [Serfati 1992], 98–103.

However, Lambert also sometimes uses the term "transcendent"! This naturally causes some confusion, which I will now try to dissipate.

To fully understand Lambert's texts, we must remember that after Leibniz, the term "transcendent" was also used in mathematics in vague meanings of the common language and was not rigorously defined. Lambert does just that. The inflections of the meaning of the word "transcendent" in the seventeenth, eighteenth and nineteenth centuries are thus an interesting problem, which is briefly developed in section XIII-D) below.

This makes for a somewhat difficult comparison with modern statements where the word has a precise mathematical sense. To continue here with Lambert, I have thus decided to avoid it temporarily and replace it (for a while) with "not algebraic". However, the latter term has two different meanings for Euler and Lambert; it is thus appropriate to compare them as follows. Euler does not give a precise definition of "algebraic" regarding numbers, so an algebraic number is actually "expressible by radicals". This is what we concluded from the interpretation of his views (see above XIII-A). It is thus the impossibility of some specific writing that involves the non-algebraic character (in other words, "inexpressible by radicals"). For Lambert, algebraic is any quantity that is the root of an algebraic equation. Not to be algebraic is thus not to be a solution of an algebraic equation (with integer coefficients). This is the modern definition (and the one that survived) of transcendent numbers.

XIII-C RECEPTION OF THE *ENCYCLOPÉDIE* (1784–1789)

One may rightly ask whether a general and recapitulative text, published in the late eighteenth century as the three volumes of the *Encyclopédie Méthodique* which are devoted to mathematics might have been capable of reviewing the definition of transcendence in a little more depth than Leibniz's contemporaries. Yet, if the *Encyclopédie Mathématique* includes in its third volume the input word "*transcendent*" (equations), it offers little definition and even creates some conceptual confusion. The text begins by defining transcendent equations as the only differential equations (which already represents a serious limitation!) that cannot be integrated algebraically. Note that this requires a preliminary demonstration of this impossibility. In these conditions, it is thus difficult to know in advance if a given differential equation is or is not transcendent. Here is the text:

> TRANSCENDENT (*Equations*) They are [...] those which do not contain, as the algebraic equations, finite quantities, but differentials or fluxions of finite quantities; it is required that these equations between the differentials must be such that they cannot be reduced to an algebraic equation; for example, the equation $dy = \dfrac{x dx}{\sqrt{a^2 + x^2}}$ which appears to be a *transcendent* equation is actually an algebraic equation [...] but the equation $dy = \dfrac{dx}{\sqrt{a^2 - x^2}}$ is a *transcendent* equation, because one cannot express, in finite terms, the integrals of each member of this equation: the equation that represents the relation between an arc of a circle and its sine is a *transcendent* equation, because Newton demonstrated that the relation could not be expressed by any finite algebraic equation; hence it follows that it can be done only by an algebraic equation with an infinite number of terms, or by a *transcendent* equation.

In modern terms, by putting $a^2+x^2=t$, the first example immediately gives $y = \sqrt{a^2+x^2} + C$. By putting x=at, the second example leads to y= Arcsin (x/a)+C, that is x=a sin (y-C). The (negative) conclusion that the *Encyclopedia* ascribes to Newton was obviously also that of Leibniz. Really, the "differential equations" of the *Encyclopedia* are simply quadratures. To consider a quadrature or a given differential equation as "transcendent", it is thus necessary to prove that it does not admit an algebraic primitive function, which, in many cases, is far from obvious.

The text then moves on to ordinary equations (i.e, non-differential):

> One usually places among the transcendent equations the exponential equations, although these equations may only contain finite quantities; but these equations are different from algebraic because they contain variable exponents, and because we can remove these variable exponents only by reducing the equation to a differential equation. For example, let y = a^x which is an exponential equation; it is necessary (...) to differentiate the equation that will give $dx = \dfrac{dy}{y}$; this is a differential and *transcendent* equation.

This example is directly inspired by Leibniz's conceptions (see *supra* IV-D5). It is nevertheless a circular reasoning: if y=a^x is a transcendent equation, it is because the differential equation $dx = \dfrac{dy}{y}$ is transcendent and this is true only because its solutions depend on y=a^x, that is on an expression that is itself assumed by the authors not to be algebraic, that is to say, transcendent.

Moreover, a transcendent curve is simply defined in the text as "a curve having no algebraic equation, but only a transcendent equation". Next comes a note on Descartes to indicate that he called such curves "mechanical" and that he "wanted to exclude them from the geometry", "unlike Leibniz and Newton". The conclusion of the article is in the purest Newtonian style (cf. *supra* XI-F, *On the philosophy of simplicity: Descartes versus Newton*). Indeed, in the construction of geometrical problems, a curve is not to be preferred to another because it is determined by a simpler equation but because it is easier to describe geometrically.

XIII-D ON THE TRANSCENDENCE OF NUMBERS:
THE WORD AND THE CONCEPTS, FROM LEGENDRE TO HILBERT

As noted above, for Leibniz the term "transcendent" does not hold a philosophical or theological connotation. We also saw that in Euler, the term is used with precision as regards the functions, but, in the case of numbers, it does not correspond to a particularly explicit concept and, in any case, not to the concept that history has retained. Lambert, who is only interested in numbers, uses the modern concept of transcendence. He also uses the word "transcendent", however in a different meaning from the one associated with the modern concept. To indicate the modern concept that he brings to light, Lambert first uses "number at random" and, concurrently, "transcendent quantity", which is however endowed with fluctuating meanings (see above XIII-B).

Regarding the numbers, the history of the crossed reception of the term and the concept of transcendence presented, after Lambert, several curious aspects. I shall not develop this point in detail here but I will send the reader, as far as the eighteenth and nineteenth centuries are concerned, to my very widely-documented study [Serfati 1992]. For results and comments over more modern or contemporary periods, the reader should also consult [Waldschmidt 1999] and [Waldschmidt 2012][36].

From Legendre to Liouville

We shall thus briefly summarize the state of the question after Lambert. For almost two centuries, the concept of "transcendent numbers" flourished but not under that name. Instead, through a circumlocution, authors used another formulation: "number that is not included in algebraic irrationals". So, at the end of the eighteenth century (1794 for the first edition), Legendre proved the irrationality of π and π^2 in an appendix to his *Eléments de géométrie* (note 4, pages 289–296, ed. 1842)[37]. Without prior explanation, he begins with a hypergeometric series expansion and makes a change of unknown function, so as to find a functional equation that he very skillfully interprets as an expansion of a continued fraction – all of this without mentioning any issue of convergence. In other work, I have analyzed his proof in modern terms[38]. Legendre *in fine* postulates the transcendence of π (in the modern sense) in these terms:

> It is likely that the number π is not even included in the algebraic irrationals, that is, it cannot be the root of an algebraic equation with a finite number of terms the coefficients of which are rational; but it seems difficult to rigorously establish this statement. (p. 289)

Somewhat later, in a number of important and famous articles, Liouville demonstrates the existence of the first transcendent numbers (in the modern sense). As noted by Stewart, it was not even clear at that time that transcendent numbers existed! This stage was therefore essential in the history of ideas. Even if Liouville operates on the objects (the "quantities"), he never uses the word "transcendent", however. Three articles between 1844 and 1851[39], both contained in the *Comptes Rendus* and in the *Journal de Liouville*, have exactly the same title:

> "Sur des classes très-étendues de quantités qui ne sont ni algébriques, ni même réductibles à des irrationnelles algébriques" (*On very extensive classes of quantities which are neither algebraic, nor even reducible to algebraic irrationals*).

Note incidentally that Liouville demonstrates his central result of the existence of the transcendent numbers by uncovering a property of the "other numbers" (i. e., the

---

36   These two historico-epistemological articles cannot by themselves account for the exceptional richness of the mathematical work of Waldschmidt on the subject of transcendence. One will find a partial bibliography in [Waldschmidt 2012], 65–67. See also [Waldschmidt 2000].

37   [Legendre 1842], *"Où l'on démontre que le rapport de la circonférence au rayon ainsi que son quarré sont des nombres irrationnels"* (*"Where we demonstrate that the ratio of the circumference to the radius, as well as its square, are irrational numbers"*).

38   See my comment in [Serfati 1992], 84–85, and a mathematical appendix on the subject (*Supplement N° 3, Sur la démonstration de Legendre de l'irrationalité de p*, idem, 177–179).

39   [Liouville 1844a], [Liouville 1844b], [Liouville 1844c].

algebraic numbers, still in the modern sense). This is a result that I have also stated in informal terms as follows[40]: algebraic numbers are "poorly surrounded by the rational numbers" (these are *Liouville's inequalities*). So, if a number happens to be well surrounded by the rational numbers, then it is transcendent (in the modern sense); for example $\sum_{n\geq 0} \dfrac{1}{10^{n!}}$, that is a numerical series whose partial sum converges much too fast.

Moreover, Liouville demonstrates that, contrary to what false intuition could suggest, these transcendent quantities are very numerous: there is actually a *very extensive class* of such numbers. He thus finds the hypothesis of Euler (see XIII-A above). The expression "extensive class" actually recovers here a very precise property, which is demonstrated, in essence only, by Liouville in his articles. The expression was later established, formally and precisely, by Cantor; namely, that the set of such transcendent numbers (in the modern sense) has the cardinality of the continuum[41]. Liouville therefore exploits perfectly the concept of transcendence but he does not use the word. Note also that Liouville's proof is *constructive*: it allows the construction of diverse transcendent numbers but it tells us nothing of the transcendence of numbers given in advance, such as the usual numbers of the analysis, for example e or π.

Hermite and Lindemann: "La transcendance est fille de l'irrationalité"[42]

At the end of the nineteenth century, through four outstanding *Compte Rendus* to the Académie des Sciences (*Sur la fonction exponentielle*[43], Hermite demonstrated the transcendence of e but he did not use the term "transcendent". The method of Hermite shows an exceptional mathematical inventiveness and is the real cornerstone in the history of the proofs of transcendence of numbers – it is the dawn of the modern era on this subject. Hermite's technique consists in approximating *simultaneously* diverse exponentials by rational fractions. The method, which has become classic in contemporary education, is based on the re-examination of a new demonstration of the irrationality of e proposed by Fourier in 1815.

Hermite himself does not give many genuine indications about his governing ideas. The 1873 article presents without prerequisites an introduction to the rational approximations in the form an analogy. Similarly, he presents his fundamental polynomial "in a very synthetic way", as Lebesgue writes[44]. Admittedly, Hermite wrote

---

40   See my comments in [Serfati 1992], 90–97, and a mathematical appendix on the subject (*Supplément No. 4 Démonstration moderne du théorème de Liouville), idem,* 180–181. See also a study of Liouville's demonstration and a presentation of Liouville's numbers in [Waldschmidt 1999], 78–84.

41   See [Serfati 1992], 107–109 and [Waldschmidt 1999], 83–84.

42   This is a favorite Waldschmidt maxim. It can be (approximately) translated as "Transcendence is an offspring of irrationality".

43   [Hermite 1873a].

44   [Lebesgue 1932], 226.

in a letter to Borchardt, "I only do what Lambert did"[45], but the relation between the methods of Hermite and Lambert is not obvious. Lebesgue's sagacity, as well as his obstinacy, was necessary to disentangle what Hermite argued about the real relation of his method with that of Lambert. In his *Leçons sur les Constructions Géométriques*[46], Lebesgue tries to genuinely understand the already mentioned issue of rational approximations for exponentials. I have elsewhere analyzed in detail his commentary[47].

According to Waldschmidt, one cannot find to this day a proof of transcendence that is fundamentally different from that of Hermite[48]. On the other hand, this method of transcendence found its basis in a proof of irrationality (that of Fourier). One thus understands the foundation of the aforesaid maxim, "*La transcendance est fille de l'irrationalité*". We can finally note that Hermite's result involves the first constant of the analysis proven to be transcendent[49].

Nine years after Hermite, Lindemann did the same with the transcendence of π (*Über die Zahl π*)[50]. His method uses essentially all of Hermite's ideas while he nevertheless added several modifications. After the preparatory work of Wantzel (1837, *Recherches sur les moyens de reconnaître si un problème de géométrie peut être résolu par la règle et le compas*)[51], the proof of the impossibility of squaring the circle was obtained[52].

So, an argument of transcendence terminated a thousand-year-old problem of quadrature.

45 [Hermite 1873c].
46 [Lebesgue 1950].
47 "Hermite et Lambert" in [Serfati 1992], 152–153.
48 Cf. [Waldschmidt 2003]: "The thesis that I would like to advocate is that there exists essentially a single method allowing to show that some numbers are transcendent: it is the one that allowed Hermite to demonstrate the transcendence of the number *e* in 1873. This method has many variants, of which we can highlight the common characteristics".
49 I have detailed the various stages of this proof in [Serfati 1992], 110–144. Here are the main stages: 1) the simultaneous approximation of n real numbers (121); 2) the problem of a functional extension of the previous numerical approximation property (121–123); 3) is this extension possible? (123–125); 4) the fundamental re-examination of the demonstration of irrationality of *e* due to Fourier (1815), which became a classic that replaced the work of Lambert (125–126); and 5) the epistemological search for a way to pass from a demonstration of irrationality to one of transcendence (126–127). I have also given a modern version of the demonstration (*Supplément N° 6: Démonstration moderne de la transcendance de e*), *idem*, 183–186. See also [Waldschmidt 1999], 84–91.
50 [Lindemann 1882]. See our study, *Lindemann: de la transcendance de π à la non quadrature du cercle,* in [Serfati 1992], 155–162. See also [Waldschmidt 1999], 91–92.
51 *Researches on the ways to recognize if a geometrical problem can be solved by ruler and compass* = [Wantzel 1837]. See Wantzel, *La règle et le compas,* in [Serfati 1992] 98–103.
52 Lindemann writes in the introduction: "Given the failure of so many attempts to square the circle with ruler and compass, the problem is generally considered to be impossible; however, it was established only the irrationality of π and π^2. The impossibility of squaring will be proved if it is shown that π cannot be the root of any algebraic equation with rational coefficients; the subject of the following is precisely to bring the demonstration".

Hilbert's "seventh problem": Euler-Hilbert conjecture

However, it was Hilbert's intervention in 1900 that assured the modern final coincidence between term and concept. In Paris, in 1900, the second Congress of Mathematicians opened. Hilbert was thirty-eight years old. He was already a well-known mathematician. On the advice of Minkowski, he proposed to the Congress twenty-three problems that he declared were promising for the future[53]. Among them, he resumed in an equivalent form Euler's conjecture (see XIII-A) above), by giving "transcendent" its modern and definitive meaning. After a reverence to Hermite and Lindemann, Hilbert expresses his seventh problem:

The expression α^β for an algebraic base α and an irrational algebraic exponent β, e.g., the number $2^{\sqrt{2}}$ or $e^\pi = i^{-2i}$, always represents a transcendent or at least an irrational number[54].

In both examples, the exponent is a quadratic irrational, a real number in the first case $\left(\sqrt{2}\right)$ and a complex number in the second (the minimal polynomial of $-2i$ is $x^2 + 4 = 0$).

In short, for Hilbert, "the irrationality of the exponent causes the transcendence of the exponential".

Actually, Hilbert resumed Euler's conjecture in a form equivalent to that in which Euler proposes it. The latter also uses the term "transcendent". But Euler (obviously!) understands it in the Eulerian sense (see above), while Hilbert understands it (of course!) in the definitively modern sense (i.e., Lambertian).

The first partial solution of the Euler-Hilbert problem, where β is a complex quadratic irrational number, was given by Gel'fond in 1919[55]. It therefore included the transcendence of e^π. Kuzmin (1930) and, independently, Siegel proved that Gel'fond's method could be extended in case β is a real quadratic irrational number, thereby establishing the transcendence of $2^{\sqrt{2}}$. In 1934, Gel'fond (see [Gel'fond 1934]) and, independently, Schneider (see [Schneider 1934)], gave a complete solution for the Euler-Hilbert problem[56].

---

53 See the full text of Hilbert's presentation at the Congress in [Browder 1976].
54 See [Browder 1976],15–16.
55 [Gel'fond 1929].
56 See commentaries on this issue in [Waldschmidt 1999].

CHAPTER XIV
COMTE AND THE PHILOSOPHY
OF THE TRANSCENDENT ANALYSIS

The first volume of the *Cours de Philosophie Positive*[1] (1830) of Auguste Comte is almost entirely devoted to the Mathematical Sciences. In eighteen lessons, the author develops new considerations for the time, often interesting, on mathematical analysis and the differential and integral calculus, but also on the calculus of variations, analytical geometry in two or three dimensions, rational mechanics, statics and dynamics. Comte knows mathematics, at first through his scientific training, and also having taught them[2]. Even if one cannot here analyse in detail a work which, since its publication, has given rise to an exceptional abundance of comments, it seemed to me useful to briefly examine its connection with the object of the present book. We shall thus focus on the conception, essentially philosophical, which Comte develops, of what he calls the *transcendent analysis* – an expression which did not however survive him – as opposed to the *ordinary analysis*[3].

The glorification of the transcendent analysis

The term 'transcendent' essentially appears in the expression 'transcendent analysis'; however, this reference is omnipresent in the book. Auguste Comte awards supported praise to it. He greatly admires this analysis by which, he says, Descartes, as innovative as he could be in his time, was exceeded. In the work, he clearly reports the considerable contribution of the works of Descartes to mathematical science, but he recognizes at the same time that these were partially made obsolete by the arrival of the transcendent analysis. Thus, about the method of the indeterminate coefficients, so successfully invented by Descartes in his *Géométrie* (AT VI, 420), he writes:

---

1   [Comte 1830].
2   A former student of the Ecole Polytechnique, he was appointed at the same school as tutor in Analysis and Mechanics (1832), then as an examiner for the entrance examination (1836).
3   In this chapter, I directly translated numerous excerpts from the book by Comte. In this case, I have simply indicated the mention (Comte, page). I have also included some of the English translation of the work by Harriet Martineau, *The Positive Philosophy of August Comte* ([Martineau 1853]). When the translation of Martineau seemed to me faithful to Comte's original, I simply indicated it by the mention (Comte, page. Transl. H.M). However, as indicated by its title, the book is a "translation and a condensation" of the *Cours de philosophie positive* of Comte. In such circumstances, however, a few 'condensations' of Martineau, although relevant, are sometimes quite distant from the French text. When I had to complete or modify Martineau's text, in order to make it compliant with the original, I indicated it by the mention (Transl. H.M, according to Comte, page) following the translation. In all cases, however, I noted 'transcendent' where Mrs Martineau noted 'transcendental'.

> This eminently analytical method is one of the most remarkable discoveries of Descartes. The invention and development of the infinitesimal calculus, for which it might be very happily substituted in some respects, has undoubtedly deprived it of some of its importance. (Comte, p. 214. Transl. H.M)

Comte's judgment in this case seems to have relatively little grounding. The method of the indeterminate coefficients is today still a very remarkable symbolic tool, whatever the mathematics domain concerned, including differential calculus. But Comte wants to glorify without reservation the transcendent analysis, which he places in the rank of the "most admirable instrument for the mathematical exploration":

> (...) the conception, highly productive, that the human mind succeeded in establishing, and which constitutes his most admirable instrument for the mathematical exploration of natural phenomena, that is the (so-called) transcendent analysis. (Comte, p. 191)

In the transcendent analysis, Comte distinguishes clearly between the differential calculus and the integral calculus. For him (see p. 238), the transcendent analysis is of use to everything: in the quadratures, the rectifications, etc. All the physico-mathematical research, he says, is there reduced in some formulas:

> (...) In virtue of this second property of the analysis, the entire System of an immense science, like geometry or mechanics, was submitted to a condensation into a small number of analytical formulas. (Comte, p. 240. Transl. H.M)

Driven by his enthusiasm, Comte even ensures that the transcendent analysis is thus the real domain of the "general and abstract processes":

> But this preliminary elaboration has been remarkably simplified by the creation of the transcendent analysis, which has thus hastened the moment at which general and abstract processes may be uniformly and exactly applied to the solution, by reducing the operation to finding the equations between auxiliary magnitudes. (Comte, p. 260. Transl. H.M)

On the philosophy of mathematics

As a coherent philosopher of mathematics, Comte wants to develop an analytical study and a rational inventory of the mathematical science at the time he wrote, and in particular, to situate the place of the transcendent analysis in the panorama of mathematical disciplines. This he carries out in the work here in question, obviously within the specific frame of his 'positive philosophy'. We shall repeat, nevertheless, that even if his terminology of 'transcendent analysis' did not survive, the work remains interesting for our subject. One of its main attractions lies certainly in the fact that, one and a half centuries after Leibniz, it reviews the history of the ideas as to the differential and integral calculus. The publication of the *Cours* is indeed situated after Comte had knowledge of the works of Euler, D'Alembert, and especially Lagrange, which he obviously knows well– he repeatedly makes a detailed analysis of the works of the latter.

One of the basic philosophical elements for Comte, present throughout the book, is what he calls the development of the 'abstract', starting from the 'concrete'. First of all, for him, is this pair of opposites: concrete mathematics *versus* abstract mathematics. Concrete mathematics is, he says, "composed only of geometry and mechanics" (p. 143). As for abstract mathematics:

> (…) it consists of what we call the calculation, taking the word in its largest extension, which encompasses from the simplest numerical operations up to the most sublime combinations of the transcendent analysis. (Comte, p. 143)

This evolution, which leads mathematical science towards a more and more abstract aftermath is, for Comte, an unavoidable necessity in the elaboration of human knowledge:

> (…) the importance and the scope necessarily predominant of abstract knowledge, of which the smallest naturally corresponds to a multitude of concrete researches, the man having no other means for the successive expansion of his intellectual capacities than in the consideration of ideas more and more abstract and yet positive. (Comte, p. 314)

About equations: formation *versus* resolution

Such a duality of the abstract/concrete, Comte believes it can be found embodied within mathematics by the opposition between the *formation* of equations and their *resolution*. This distinction, to which he lends a philosophical nature, is a key point for understanding the thoughts of the author. Comte then introduces the division between direct and indirect functions, a categorization that runs through the book. So he writes:

> We now see that the *Calculus of functions*, or *Algebra*, must consist of two distinct fundamental branches. The one has for its object the resolution of equations when they are directly established between the magnitudes in question: the other, setting out from equations (generally much more easy to form) between quantities indirectly connected with those of the problem, has to deduce, by invariable analytical procedures, the corresponding equations between the direct magnitudes in question – bringing the problem within the domain of the preceding calculus. (Transl. H.M, according to Comte, p. 195)

The *resolution* of equations (by ordinary analysis, that is the calculation of the direct functions) is therefore strictly different from their *formation* (by the transcendent analysis, that is to say, by the calculation of the indirect functions):

> The first calculation, most of the time, bears the name of *ordinary analysis*, or *algebra*; the second constitutes what we call the *transcendent* analysis, which is indicated by the diverse names of *infinitesimal calculus, calculus of fluxions and fluent quantities*, of *vanishing elements*, etc. according to the point of view under which it was conceived. To rule out any foreign consideration, I suggest naming it the *calculation of the indirect functions*, by giving to the ordinary analysis the title of *calculation of the direct functions*. (Comte, p. 191)

Regarding the differential calculus (here considered as opposed to the integral calculus), Comte considers that the differential elements are "auxiliary quantities" from which the mathematician "goes back to the primitive quantities"; they thus serve to *form* the equations, and fall within the calculation of the indirect functions.

With regard to Newton, Comte will return to the status of some "auxiliary magnitudes", which are certain "limits of ratios" between the primitive quantities:

> Under this point of view, the general spirit of the transcendent analysis consists in introducing as auxiliary – instead of the primitive quantities or jointly with them, to facilitate the establishment of the equations – the limits of the ratios of the simultaneous increases of these quantities, or, in other words, the last reasons of these increases, limits (…). (Comte, p. 248–249)

One can summarize as follows Comte's conclusions regarding the differential analysis: starting from a 'concrete' problem, differential calculus is used to form equations which are then solved, either by ordinary analysis (or, he says, algebra) or by integral calculus. In short, the differential calculus is primarily used to form equations between ordinary entities, starting from the differential quantities. Then, the calculation serves to solve, if it is possible, these equations. Such a position is constant. Accordingly, the transcendent analysis, which is logically the first, should have been the first to be taught and practised. But this, Comte says, is contrary to standard usage:

> In this respect, and according to the rational order of the ideas, transcendent analysis appears as being inevitably the first one, since it has the general aim of facilitating the establishment of the equations; and this obviously has to precede the actual resolution of these equations, which is the object of the ordinary analysis. However, although it is eminently important to conceive in this way the real concatenation of these two analyses, it is nevertheless appropriate, in accordance with the constant practice, to study the transcendent analysis only after the ordinary analysis. (Comte, p. 194)

Note also (see below) that, for Comte, this calculation can often be completed only through the intervention of the integral calculus, which, he will also say, is still very imperfect.

Leibniz, Newton, Lagrange: the trio of interpreters

For Comte, there is an objective scientific content, namely the transcendent analysis. Nevertheless, historically this was subject to various successive subjective interpretations, mainly three in number, from Leibniz, Newton, Lagrange:

> This analysis [namely, that of the philosophical character specific to the transcendent analysis] was, as we know, presented by the mathematicians under several really distinct views, although always necessarily equivalent and leading to identical results. We can reduce them to three main ones, those of Leibniz, Newton and Lagrange, of which all others are but secondary modifications. In the present state of the science, each of these three general conceptions offers essential advantages which belong to it exclusively – and we did not succeed until now in combining all these diverse characteristic qualities into a single method (Comte, p. 225)

The contributions of our three protagonists will then be widely commented upon in the work, both historically and philosophically. There are thus, for Comte, three names for the same differential calculus, associated with our three authors: infinitely small quantities (Leibniz), fluxions (Newton), and derivatives (Lagrange):

> But, at present, the only auxiliary quantities habitually substituted for the primitive quantities in *transcendent analysis* are what are called – Ist, *infinitely small elements*, the *differentials* of different orders of those quantities, if we conceive of this analysis in the manner of Leibnitz: or 2nd, the *fluxions*, the *limits* of the ratios of the simultaneous increments of the primitive quantities, compared with one another; or, more briefly, the prime and ultimate ratios of these

increments, if we adopt the conception of Newton: or. 3rd, the derivatives, properly so-called, of these quantities; that is, the coefficients of the different terms of their respective increments, according to the conception of Lagrange. (Comte, p. 193. Transl H.M)

At the symbolic level, Comte also distinguishes the specific representations of the three protagonists (p. 262):

$$(dy/dx) \text{ (Leibniz)} \rightarrow \lim (\Delta y / \Delta x) \text{ (Newton)} \rightarrow f'(x) \text{ (Lagrange)}$$

Let us repeat: the conception of Comte is that there exists an underlying *objective* content, that of the transcendent analysis, which is the same for three *subjective* interpretations. Note that the introduction of his very general terminology of direct and indirect functions – which did not survive him – consolidates his conception. Thereby, he is indeed permitted not to specify which of the three interpretations of the transcendent analysis with which he deals:

> This is what determined me to regularly define the transcendent analysis, as the calculation of indirect functions, to mark its real philosophical character, by discarding any discussion about the most appropriate way to develop it and apply it. (Comte, p. 259)

And he insists, for him, all three calculations are "necessarily identical":

> These three main ways to consider our present transcendent analysis, and all the others less distinctly defined which have been successively proposed, are, by their nature, necessarily identical, either in calculation or in the application. (Comte, p. 193)

Such a position of principle, however, may appear somewhat Platonist in that it ignores the epistemological question of the diverse modes of genesis of the concerned concepts.

For a reasoned history

Comte then proposes a reasoned history of "the successive formation of the transcendent analysis", starting with the Ancients, then continuing with Fermat, to whom he awards, quite rightly, the title of precursor:

> The first germ of the infinitesimal method (which can be conceived of independently of the Calculus) may be recognized in the old Greek Method of Exhaustions. Employed to pass from the properties of straight lines to those of curves. (Comte, p. 227)

But, says Comte pertinently, the process of the Ancients was not really a method: "they had no rational and general means for the determination of these limits!" And it was not until the seventeenth century, at first with Fermat, that the first germs of a general calculation method appear; for his part, Lagrange also had rightly recognized this point:

> For the ancients had no logical and general means for the determination of these limits, which was the chief difficulty of the question. The task remaining for modem geometers was to generalize the conception of the ancients, and, considering it in an abstract manner, to reduce it to a system of calculation, which was impossible to them. Lagrange justly ascribes to the great geometer Fermat the first idea in this new direction. Fermat may be regarded as having initiated the direct formation of transcendent analysis by his method for the determination of maxima and minima. (Transl. H.M., according to Comte, p. 228)

Leibniz, the authentic creator of the transcendence.
"The loftiest idea ever yet attained by the human mind"

So, there were certainly three main protagonists. But, says Comte in his seventh lesson[4], the only authentic creator of the transcendent analysis was Leibniz who, by relying initially on the works of Wallis and Brouncker, came to exceed Fermat. Comte then resumes the history of the differential calculus. At first he grants to Descartes his "great fundamental idea", namely the "general analytical representation of natural phenomena" (p. 236). But the fundamental design of the transcendent analysis, through which Descartes was to be supplemented, was unquestionably the fact of Leibniz:

> But, although Fermat had first developed this analysis in a truly abstract way, it was still far from being regularly formed as a general and different calculation, with its own notation, and especially disengaged from superfluous consideration of the terms, which turned out to no longer be considered within the analysis of Fermat, although they nonetheless greatly complicated all the operations by their presence. It was so successfully executed by Leibniz half a century later, after some intermediate modifications to Fermat's ideas brought by Wallis, and above all by Barrow, and thus he was the real creator of the transcendent analysis as we use it today. (Comte, p. 229)

Comte continues, by crediting Leibniz with the "loftiest idea ever yet attained by the human mind", with this real song of glory addressed to him:

> Such a mere hint as this of the varied applications of this method may give some idea of the vast scope of the conception of transcendent analysis, as formed by Leibnitz. It is, beyond all questions, the loftiest idea ever yet attained by the human mind. It is clear that this conception was necessary to complete the basis of mathematical science, by enabling us to establish, in a broad and practical manner, the relation of the concrete to the abstract. In this respect, we must regard it as the necessary complement of the great fundamental idea of Descartes on the general analytical representation of natural phenomena; an idea which could not be duly estimated or put to use till after the formation of the infinitesimal analysis. (Comte p. 236. Transl. H.M.)

Comte, although he attributes to Leibniz the whole invention of the new analysis, will however further note some reservations about the rigor of the process of his creation:

> Leibniz (…) had himself submitted an entirely erroneous explanation, by saying that he treated the infinitely small quantities as incomparable elements, and that he neglected them vis-à-vis the finite quantities, as the grains of sand compared to the sea. It was a consideration which would have completely denatured his analysis, by reducing it to nothing more than a simple calculation of approximation, which, in this respect, would be radically vicious, since it would be impossible to predict, as a general rule, to what extent the successive operations can enlarge these early errors, the increase of which could even become obviously arbitrary. Leibniz had thus glimpsed, only in an extremely confused way, the true rational foundations of the analysis which he had created. (Comte, p. 242)

This critique of Comte, however, seems to me somewhat misplaced. Leibniz, once past the time of the first discoveries in his youth, had indeed perfectly understood the importance of the upcoming theory he had brought to light. More importantly,

---

4   *Tableau général du calcul des fonctions indirectes.*

he had perfectly made the distinctions between the "approximations" and "all the approximations", and also "the true value" (see example above in Chapter VII). In an excellent article ("Le Continu chez Leibniz"[5]) Breger fully overcomes the criticisms on the alleged lack of rigor of Leibniz.

Newton: an effort to rationalize

Comte then passes to the case of Newton. He first explains that the idea of the discovery of Leibniz was indeed 'ripe' at the moment, and that, at the same time, other scholars were interested independently, but understanding it however under other perspectives:

> This eminent discovery was so ripe, as all great conceptions are at the hour of their advent, that Newton had at the same time, or rather earlier, discovered a method exactly equivalent, regarding the analysis from a different point of view, much more logical in itself, but less adapted than that of Leibnitz to give all practicable extent and facility to the fundamental method. (Comte, p. 229. Transl. H. M.)

The author then returns to Newton. Comte, as he himself is a good mathematician, is quite comfortable to understand and explain his practices:

> Newton offered his conception under several different forms in succession, that which is now most commonly adopted, at least on the Continent, was called by himself, sometimes the Method of *prime and ultimate ratio*, sometimes the *Method of Limits*, by which last term it is now usually known. (Comte, p. 248. Transl. H. M.)

For example, as regards the determination of the tangent to a plane curve, one obtains, by following Newton:

> (…) for the trigonometric tangent of the angle which is made by the secant line with the X-axis, $t = \frac{\Delta y}{\Delta x}$, from which, by taking the limits, we shall deduce, with respect to the tangent itself, this general formula of transcendent analysis

$$t = \lim \frac{\Delta y}{\Delta x} \quad (1)$$

> (…) according to which the calculation of the indirect functions will teach in every particular case, when the equation of the curve will be given, how to deduce the relation between t and x, by eliminating the auxiliary quantities introduced. (Comte, p. 250)

Further on, Comte accurately explains the definitions in Newton of the *fluxions* and the *fluent* quantities:

> The velocity with which each one will have been described will be called the fluxion of that quantity, which inversely would have been called its fluent. Henceforth, the transcendent analysis will, according to this conception, consist in forming directly the equations between the fluxions of the proposed quantities, to deduce from them afterwards, by a special Calculus, the equations between the fluents themselves. (Comte p. 253. Transl. H.M)

5   [Breger 2016], 127–135. See in particular "La justification", 132–133.

Accordingly, the conception of Newton is apparently more rigorous than that of Leibniz:

> Turning to the conception of Newton, it is clear that, by its nature, it is safe from the funda-
> mental logical objections which are caused by Leibniz's method. The *notion of limits* is indeed
> remarkable by its clarity and its adequacy. In the transcendent analysis presented in this way,
> the equations are considered to be exact from the beginning and the general rules of reasoning
> are as constantly observed as in the ordinary analysis. (Comte, p. 256)

This conception is not, however, exempt from a severe ontological weakness, namely that of not being able to conceive "the increments of the quantities considered separately and in themselves, no more than in their ratios, but only within the limits of these ratios". Comte added that this deficiency:

> (...) significantly slows down the course of the mind for the formation of the auxiliary equa-
> tions. We can even say that it hinders the purely analytical transformations. (Comte, p. 266)

And a little further, still on the subject:

> I must not neglect in this regard to observe that many geometers of the continent, by adopting
> (as being more rational) Newton's method to serve as a basis for the transcendent analysis, have
> partially disguised this inferiority by a grave inconsequence, which consists in applying to this
> method the notation imagined by Leibniz for the infinitesimal method, which is really appropri-
> ate only for the latter. (Comte, p. 266)

Lagrange, and the promotion of "abstraction"

As seen above, Comte considered that each of the three conceptions of our authors had its advantages and its disadvantages. It is however in that of Lagrange that he places, unquestionably, what will be, in the future, the theory of transcendence. It is Lagrange's scheme that will eventually overcome, he says. And when it will have triumphed, the rest will have only an historic interest:

> We referred (p. 44) in a former section to the views of the transcendent analysis presented by
> Leibnitz, Newton, and Lagrange. We shall see that each conception has views, has advantages
> of its own; that all are finally equivalent, and that no method has yet been found which unites
> their respective characteristics. Whenever the combination takes place, it will probably be by
> some method founded on the conception of Lagrange. The other two will then offer only an
> historical interest. (Comte p. 225. Transl. H.M)

Further on he still insists that the perfect character of the interpretation of the transcendent analysis can be found only in Lagrange:

> This perfect unity of analysis, and a purely abstract character in its fundamental ideas, are
> found in the conception of Lagrange, and there alone. It is therefore the most philosophical of
> all. Discarding every heterogeneous consideration, Lagrange reduced all transcendent analysis
> to its proper character, – that of presenting a very extensive class of analytical transformations
> (Comte p. 268. Transl. H.M).

Here, Comte is prophetic. It is indeed Lagrange's conception that triumphed today; it is indisputably the source of modern differential calculus. To construct from it fruitful successive extensions, the mathematicians of our time had nevertheless to find an adequate frame of extension, mainly that of normed vector spaces (especially Banach spaces)[6], and become free within this framework from the concept of *quotient*, which yet appeared, in one form or another, consubstantial with the definition of the derivative of a function[7].

The design of Lagrange is then, in turn, finely analysed in the paper. For Comte, the calculation of derived functions which Lagrange introduced behaves like "a very considerable extension of the ordinary analysis". One completes the latter by the practice of derivation and of successive derivations – this, we add, is obviously not a small contribution! Comte analyses the conception of Lagrange in these terms:

> The transcendent analysis is then only a simple, but very considerable extension of ordinary analysis. It has long been a common practice with geometers to introduce, in analytical investigations, in the place of the magnitudes in question, their different powers, or their logarithms, or their sines, etc., in order to simplify the equations, and even to obtain them more easily. Successive derivation is a general artifice of the same nature, only of greater extent, and commanding, in consequence, much more important resources for this common object. (Comte, p. 255 Transl. H.M)[8]

The constant importance given by Comte to the elaboration of the abstraction in mathematics has already been noted above. That he recognizes, of course, in Lagrange's conceptions. However, if he insists on that issue, he will also express some reservations on the latter: this time he considers these conceptions as too much abstracted (!):

> This is the weak side of Lagrange's great idea. We have not yet become able to lay hold of its precise advantages, in an abstract manner, and without recurrence to the other conceptions of the transcendent analysis. These advantages can be established only in the separate consideration of each principal question; and this verification becomes laborious in the treatment of a complex problem. (Comte, p. 256. Transl. H.M)

6   See a modern presentation of the issue in [Cartan 1967].

7   How indeed to define the derivative $f'(a)$ of a function $f: U \to F$, at a point a of U, if E and F are any two normed vector spaces (they are not necessarily finite-dimensional) and U an open subset of E? Historically, the question had arisen very early, and in the simplest cases where f is for example a real function of two real variables ($f: IR^2 \to IR$). Even in this case, the research would have seemed vain, any attempted solution stumbling over the fact that the natural extension from the ordinary derivation ($f: IR \to IR$) cannot spontaneously be made: the usual formula of definition of the derivative uses organically a quotient, and one cannot envisage a quotient of vectors. The modern definition of a *differentiable function* bypasses (easily!) this difficulty (see our commentary in [Serfati 2005], 374).

8   In his book, *Théorie des fonctions analytiques* (1797), Lagrange glorifies the use of derivatives, which he will use abundantly in all the work (see [Fraser 2005]). The book's subtitle explains to what extent Lagrange wants to distance his new theory from the interpretations of Leibniz and Newton: "Theory of analytical functions containing the principles of the differential calculus disengaged from all consideration of infinitesimals, vanishing limits or fluxions and reduced to the algebraic analysis of finite quantities".

Differential calculus and geometry in two and three dimensions

In a set of conclusions contained in the thirteenth and fourteenth lessons, Comte explains then, for the use of the contemporary reader, some of the geometrical applications of the differential calculus, in dimensions two and three. First of all, obviously, this calculation is intended for the determination of the tangents to plane curves. This application had originally been the central motivation of the young Leibniz to promote his calculus. Thus Comte writes:

> I therefore turn immediately to the consideration of the theories of general geometry in two dimensions which can be completely established only by means of the transcendent analysis. The first one and the simplest of them consists in the determination of the tangents to plane curves. (Comte, p. 486)

The tangent is however, as we say today, a first-order differential element. Comte then pursues his inventory of the differential problems associated with the plane curves, with questions concerning some of the second-order differential elements, such as the curvature and the osculating circle – these were questions that Huygens and Leibniz had both studied:

> A second fundamental problem presented by the general study of curves, and of which the complete solution requires a more extensive use of the transcendent analysis, is the important issue of measuring the curvature of the curves by means of the osculating circle at each point, whose discovery would alone suffice to immortalize the name of the great Huygens. (Comte, p. 491)

In this context, Comte thematises the research of plane curves determined by a differential second-order condition, for example, with respect to the curvature or the osculating circle at the current point of the curve. This process, which leads, as it is known, to a differential equation of the second order, generalizes naturally to the second order the 'Inverse Method of Tangents', which had been such an important support for Leibniz (See above chapter VI, *Quadratures, the inverse method of tangents, and transcendence*). Comte analyses pertinently:

> It would be the same all the times when we would have to determine a curve according to an arbitrary property of its radius of curvature. This type of question is exactly similar to the simpler problems which establish what, in the origin of the transcendent analysis, we called the Inverse Method of Tangents, where it was proposed to determine a curve by a given property of its tangent at any point. (Comte, p. 497)

He then comes to glorify once again Lagrange, who, he says, has succeeded, by his theory of contacts, improving the previous conclusions (i. e. those of Leibniz and Huygens) on the curvature:

> I have considered so far the theory of the curvature of curves following the spirit of the infinitesimal method *per se*, which fits indeed more simply than any other one to any research of this kind. The conception of Lagrange, in connection with the transcendent analysis, presented, especially by its nature, special great difficulties for the direct solution of such a question, as I have already remarked in the sixth lesson. But these difficulties have fortunately so aroused the genius of Lagrange that they led him to the formation of the general theory of contacts, of which the former theory of the osculating circle is nothing more than a very simple particular case. (Comte, p. 499)

In his fourteenth lesson[9], Comte exposes the case of the *differential* problems (which he distinguishes from *metric* problems, amenable to the 'ordinary analysis') in three dimensions, particularly those of the surfaces within the usual space:

> The study of surfaces consists of a series of general questions similar to those specified in the previous lesson about the lines. It is unnecessary to consider those that depend only on ordinary analysis separately, as they are resolved by essentially similar methods; either it is a question of knowing the number of points required in fully determining a surface, or to research the centres, or else that we ask for the precise conditions of the similitude between two surfaces of the same kind, etc. There is no analytical difference other than considering equations with three variables instead of equations in two variables. I therefore turn immediately to questions that require the use of the transcendent analysis, focusing only on the new considerations they present with regard to surfaces. The first general theory is that of the tangent planes (...). (Comte, p. 511)

Differential calculus and integral calculus

What precedes concerned mainly the differential calculus, which constitutes the part, the first and most of the time inescapable, of any method of transcendent analysis. Having first distinguished carefully between infinitesimal *method* and infinitesimal *calculus* (i.e. differential), Comte points out about the latter:

> As the ideas are usually vague in this respect, I believe I have to explain (...) and show that, in every application of the transcendent analysis, a first direct and necessary part is constantly assigned to the differential calculus. (Comte, p. 276–277)

Further on, however, Comte comes naturally to the integral calculus, which is such, he says, that the utility peculiar to the differential calculus does not exist independently of it. The differential calculus can certainly form equations, but the mathematician cannot, usually, resolve those, except by an application of the integral calculus, which is so, he says, "the only directly indispensable". This conception, directly derived from Leibnizian harmony[10], is here developed, beyond even Leibniz's conceptions, to its ultimate consequences.

> It may not be apparent to all minds what can be the proper utility of the differential calculus, independently of this necessary connection with the integral calculus, which seems as if it must be in itself the only directly indispensable one; in fact, the elimination of the infinitesimals or the derivatives, introduced as auxiliaries, being the final object of the calculus of indirect functions. (Comte, p. 275. Transl. H.M).

Comte insists at length on this point: in many instances, the elimination of the infinitesimals (which is so necessary!) can be administered only by the integral calculus:

> According to these expressions, the differential general formula described above does not contain more than ds and dx; the elimination of the infinitesimals, once arrived at this point, can be completed only by the integral calculus. Such is the general function of the differential calculus

---

9   *De la géométrie générale à trois dimensions.*
10   See above, Chapter VIII, *On Harmony in Leibniz.*

necessarily appropriate to the complete solution of the questions which require the use of the transcendent analysis. (Comte, p. 278–279)

Such is thus, as says Comte, "the general function" of the integral calculus. Further on, he will nevertheless have to resign himself to comment and regret its "state of extreme imperfection":

> As for the rest, though I was bound to exhibit in my summary exposition the state of extreme imperfection in which the integral calculus still remains, it would be entertaining a false idea of the general resources of the transcendent analysis to attach too much importance to this consideration. As in ordinary analysis, we find here that a very small amount of fundamental knowledge respecting the resolution of equations is of inestimable use. (Comte, p. 313, Transl. H.M).

One can grant to Comte that he is partly right in his nuanced pessimism. Admittedly, the integration techniques allow calculation of only a small number of primitives, and integration of but a small number of differential equations, or partial derivatives equations[11]. However, other methods can supplement this imperfection of the integral calculus, as for example the definition and study of new functions *ad hoc*, in the case of the primitive functions; either, in the case of differential equations, a systematic development of approximation methods, allowing the graphical representations of the bundles of integral curves to be drawn with high accuracy. Of this situation, Comte proposes however a more philosophical, slightly too short explanation in my opinion. It is due, he says, to the fact that the transcendent analysis is "still much too near its birth". Thus, we have not yet had enough time to know what it will become:

> The transcendent analysis is still much too near its birth; especially, there has been too little time for it to be designed in a truly rational way, so that we can have a fair idea of what it could become one day. But whatever our legitimate expectations, let us not forget to consider above all the limitations of our intellectual constitution – even if they are not susceptible to a precise determination, they have nevertheless an indisputable reality. (Comte, p. 312)

Comte naturally feels the necessity of commenting, again 'rationally', on the articulation between both aspects, differential and integral, of the transcendent analysis. He proposes for this purpose a general classification of the problems according to three topics, all depending on the transcendent analysis, according to whether or not they fall within the differential calculus and / or the integral calculus:

> Thus, all questions to which the analysis is applicable are contained in three classes. The first class comprehends the problems, which may be resolved by the differential calculus alone. The second, those which may be resolved by the integral calculus alone. These are only exceptional; the third constituting the normal case; that in which the differential and integral calculus have each a distinct and necessary part in the solution of problems. (Comte, p. 281, Transl. H.M)

11　In his seventeenth lesson devoted to the dynamics, Comte also notes (p. 653): "Considering the more general case where the primitive definition of the proposed movement would only be given by an equation between the four variables, the analytical problem will consist in the integration of a differential equation of the second order relative to the function, which can be frequently impracticable, considering the extreme present imperfection of the integral calculus".

On the comparison of the interpretations

The three points of view, of Leibniz, Newton, Lagrange, are, as we have seen, complementary for Comte. Each has its specific advantages and disadvantages, which Comte, often with much penetration, attempts to describe. First, Leibniz is considered by Comte to be the genuine inventor of transcendence. Such a conclusion agrees exactly with ours in the present work. Lagrange, says Comte, has himself paid tribute to Leibniz:

> Lagrange himself, after having reconstructed the analysis on a new basis, rendered a candid and decisive homage to the conception of Leibnitz, by employing it exclusively in the whole System of his 'Analytical Mechanics.' (Comte p. 263/64. Transl. H.M)

However, for Comte, dominated by positivism, Leibniz's design is somewhat insufficient[12] in its foundation, because it carries traces of the metaphysics of his founder. He continues as follows:

> Yet are we obliged to admit, with Lagrange, that the conception of Leibnitz is radically vicious in its logical relations. He himself declared the notion of infinitely small quantities to be a false idea: and it is in fact impossible to conceive of them clearly, though we may sometimes fancy that we do (…)[13] This false idea bears, to my mind, the characteristic impress of the metaphysical age of its birth and tendencies of its originator. (Comte p. 263/64. Transl. H.M).

In other words, for Comte, Leibniz certainly had the exact intuition of what was necessary to identify and invent, and his method is as applicable as intuitive, but its foundation is not rigorous.

On the other hand, the position of Lagrange is for Comte rigorous but, at the moment when he writes it is very difficult to implement, so much it is 'abstract'. The conception of Newton is intermediate between those of the two other protagonists. So Comte writes, about the three interpretations:

> As for their relative value, we shall see that Leibniz's conception has so far got an indisputable superiority in its use, but that its logical character is eminently vicious; while Lagrange's design, admirable (…) still presents, in the applications, grave inconveniences, by slowing down the progress of the intelligence (…); the conception of Newton holds more or less the middle under these diverse aspects, being less fast, but more rational than that of Leibniz, less philosophical, but more applicable than that of Lagrange. (Comte, p. 193)

In his sixteenth lesson[14], he will also add that the method of Lagrange, so rational as it may be, is difficult and abstract, and still much too new to become "as popular" as that of Descartes:

> But these considerations may be obviously only temporary; the main embarrassments which are caused by the use of the conception of Lagrange really having no other essential cause than its novelty. Such a method is certainly not indefinitely intended for the exclusive use of a

---

12   In Comte's very specific terminology, it is "vicious" (*vicieuse*).
13   A sentence of the original text was not translated by H. Martineau. By restoring it, I obtain: "The transcendent analysis designed in such a way presents, for me, this great philosophical imperfection, of being still essentially based on such metaphysical principles, from which the human mind had so much effort to release all its positive theories ".
14   *Vue générale de la statique.*

very small number of geometers, the only ones who have a familiar enough knowledge of it to suitably use the admirable properties which characterize it: it certainly has to become later as popular in the mathematical world as the vast geometrical design of Descartes and this general progress would most probably be already almost carried out if the fundamental notions of the transcendent analysis were more universally widespread. (Comte, p. 611)

In a renewed praise of the inventive genius of Leibniz, Comte still regrets that even after Lagrange, the transcendent calculation is far from being as extensive and general than Leibniz had planned it:

> So, the transcendent calculation, considered separately from its applications, is far from offering within this method the extent and the generality that Leibniz's design had printed in it. (Comte, p. 266–267)

The contributions of D'Alembert and Lagrange.
On partial derivative equations

What is for Comte interesting and new at his time in transcendent analysis, compared with the seventeenth century, i. e. with regard to Leibniz and Newton, is mainly due to the works of D'Alembert and Lagrange. Thus, he mentions, quite naturally, the extension of the differential calculus to the case of functions of several real variables[15], with the introduction of total differentials and partial derivatives:

> Moreover, one case is generally deduced from the other one, by means of a very simple invariable principle, which consists of looking at the total differential of a function according to the simultaneous increases of the diverse independent variables that it contains, as the sum of the partial differentials which would produce the separated increase of each variable successively, if all the others were constant. It is necessary, moreover, to carefully notice on this matter a new notion, which introduces, in the system of the transcendent analysis, the distinction between the functions in a single variable and in several: it is the consideration of these diverse special derivative functions, relative to each variable separately, and the number of which increases more and more, as the order of derivation rises. (Comte, p. 285)

Also note the thematisation, rather new for the time, of the technique of changes of variable to integrate differential equations – such a method is now the daily bread in research and education:

> [such a method] has for its object the constant transformation of derived functions, as the result of determined changes of independent variables; from there results the possibility of bringing back to new variables all the general differential formulas originally established in the transcendent analysis for others (...) it can be considered as increasing the fundamental resources, by allowing to choose, in order to form at first more easily the differential equations, the system of independent variables which will appear the most advantageous, although it is not to be maintained later. (Comte, p. 289)

---

15 At the epistemological level, the handling of the differential calculus for real functions of several real variables is significantly different from that of the functions of a single variable. And the content of the present section raises many difficulties with the first-year students! For an overview in modern terms, see *Différentiabilité. Fonctions de plusieurs variables réelles*, in [Serfati 1995a], 171–217.

Finally, Comte also proposes a reflection on the calculation of partial differential equations, which, he says, was created by D'Alembert. It is, for him, a "higher branch of transcendent analysis". Comte clearly emphasizes the difficulty of this new theory, which is "still entirely in its infancy". Indeed, the resolution of partial differential equations, even very simple, uses processes of a different nature than those used in ordinary differential equations. Comte also describes, pertinently, the profoundly new mode of *dependence with respect to the data* of the solutions of partial differential equations: for the first order, for example, these solutions indeed depend on an arbitrary *function* and no longer on an arbitrary *constant*:

> Thence results the most difficult, and also the most extended branch of the integral calculus, which is commonly called the Integral Calculus of partial differences, created by D'Alembert, in which, as Lagrange truly perceived, geometers should have recognized a new calculus, the philosophical character of which has not yet been precisely decided[16]. This higher branch of transcendent analysis is still entirely in its infancy. In the very simplest case, we cannot completely reduce the integration to that of the ordinary differential equations. (Comte, p. 296. Transl. H.M)

Comte then returns on the specificities, already mentioned above, of the functions of several real variables:

> In the section on the Integral Calculus, I noticed D'Alembert's creation of the Calculus of partial differences, into which Lagrange recognized a new calculus (...) the ordinary new elementary idea in transcendent analysis, – the notion of two kinds of increments, distinct and independent of each other, which a function of two variables may receive in virtue of the change of each variable separately, – seems to me to establish a natural and necessary transition between the common infinitesimal calculus and the calculus of variations. D'Alembert's view appears to me to approximate, by its nature, very nearly to that which serves as a genera basis for the Method of Variations. (Comte p. 333, Transl. H.M)

The calculus of variations, and the 'hypertranscendent' analysis

For Comte, there is another application of transcendent analysis, namely the calculus of variations, which involves both the differential calculus and the integral calculus, and to which Lagrange, following Euler, brought a decisive contribution. At first glance, this is what we today call problems of extrema (maximum or minimum) of a functional (that is, a function from a vector space into its underlying scalar field – see footnote below). The calculus of variations is now a fully-fledged discipline of mathematical analysis[17].

---

16 A sentence of the original text was not translated by H. Martineau. By restoring it, we obtain: "A very striking difference between this case and that of the equations in a single independent variable consists, as I observed it above, in the arbitrary functions which replace the simple arbitrary constants to give to the corresponding integrals all the suitable generality".

17 We thus briefly explain the problem in modern terms: f will designate an **arbitrary** real function of one real variable (the simplest case), of differentiability class C^2 (that is to say twice continuously derivable) on an open set U of IR, which has to satisfy some constraints; it is thus an element of a vector space E of real functions of one variable.

In his eighth lesson[18], Comte exposes the character, for him indisputably philosophical, of the calculus of variations. He indeed proposes (p. 315) "to capture more
easily the philosophical character of the method of variations", and thus writes:

> (...) and it is therefore necessary to indicate briefly the special nature of the problems which
> have given rise to this hyper-transcendent analysis (Comte, p. 315, Transl. H.M).

So, it will be a question for him of an "hypertranscendent" analysis. This qualifier
is without any relation with Huygens' considerations about "supertranscendence"
(see above chap. XII). Comte pursues by a relevant history on the brachystochrone
(p. 318), the geodesics on a surface (p. 327–328), and also the isoperimetric issues
(p. 442)[19]. These three problems of the seventeenth century were at the heart of the
elaboration of variational methods:

> The really decisive impulse in this respect could come only from one of the geometers occupied
> on the continent with developing and applying the genuine infinitesimal method. It is what
> made, in 1695, John Bernoulli, by offering the famous problem of the brachystochrone, which
> suggested such a long series of similar questions. (Comte, p. 318)

And further on:

> Such is, by overview, the general way by which the method of the variations applies to all the
> diverse questions that compose the so-called theory of isoperimeters. One probably will have
> noticed in this brief exposure to what extent was used, by this new analysis, the second funda-
> mental property of the transcendent analysis (...). (Comte, p. 329)

Examining the method of resolution of the variational problems, Comte explains
that Lagrange succeeded in reducing a problem of calculus of variations to a differential calculus[20]:

> On the other hand, let K a **given** real function, of three real variables, of class C^2 on an open
> subset W of IR^3. And finally, let a and b two elements of U such that $[a,b] \subset U$, and also such
> that, for every $x \in [a,b]$: $(x, f(x), f'(x)) \in W$.
>
> Then one considers the integral $J(f) = \int_a^b K\big(x, f(x), f'(x)\big)\,dx$ (f' denotes the derivative of f).
>
> J is called a **functional** on E. The 'variable' is thus here a function (namely, f).
> The object of the calculus of variations is the search for the *extremal* functions of J, that is the
> functions f (if any) that ensure that the functional J(f) reaches a maximum or minimum value
> on E (we mean a local extrema), or the *stationary* functions of J, that is those such that the rate
> of variation associated of the functional is zero.

18 *Considérations générales sur le calcul des variations.*
19 *The basic isoperimetric problem* dates back to antiquity: among all the plane closed curves the
    perimeter of which is fixed, which is the curve (if any) that maximizes the area of the region of
    the plane that it limits? One can show that the issue is equivalent to the following: among all
    the plane curves which limit a region of some given area, which is the curve (if any), of the
    smallest perimeter?
20 This is quite relevant: with the notations above (footnote 16), *the equation, said of Euler*

    Lagrange, $\dfrac{\partial K}{\partial f} - \dfrac{d}{dx}\left(\dfrac{\partial K}{\partial f'}\right) = 0$ is indeed a necessary, but not sufficient condition on f so that this

    one is an extremal function of the functional J. We so obtain a differential equation of the se-
    cond order with respect to f.

The constant purpose of this first analytical elaboration, in the exposure of which I have not to enter here in any way, is to reach the differential equations themselves, which is always possible, and by there the issue comes back within the field of ordinary transcendent analysis; this finishes the solution, at least by reducing it to pure algebra, if we know how to perform the integration. The general objective, specific to the method of variations, is to operate this transformation, for which Lagrange has established simple, invariable rules, and always with a guaranteed success. (Comte, p. 326)

In a passionate conclusion, Comte returns to the calculus of variations, considered again in the register of the 'hypertranscendence':

This Method being only an immense extension of the general transcendent analysis (…) Lagrange invented the calculus of variations in accordance with the infinitesimal conception, properly so called, and even some time before he undertook the general reconstruction of the transcendent analysis. (Comte p. 331, Transl. H.M)

On the philosophy of the hypertranscendence.
Singular and multiple. Variables and functions.

Having pertinently detailed every aspect of the transcendent analysis, Comte, as a positivist epistemologist, therefore rightly looks for the possible extensions of this one. Such is the genuine philosophical source of the hypertranscendence. As we have just seen, he found two such extensions, which occur naturally, following the works of Lagrange and D'Alembert. We shall summarize his thoughts, by starting from Lagrange's symbolism:

$$G'(x)$$

In this symbolism, x is the sign of a variable, G that of a function, and $G'(x)$ denotes the value of the derivative of G at the point x.

A first extension is constructed by replacing in the symbolism $G'(x)$ *the singular by the multiple*: to a single variable x, one substitutes several variables $x_1, \ldots x_n$. Such a substitution generates the concept of partial derivatives $\dfrac{\partial G}{\partial x_1}, \ldots, \dfrac{\partial G}{\partial x_n}$.

A second extension was constructed (independently of the previous one) by replacing, in the symbolism $G'(x)$, *the variable of sign x by a function of sign f*, to obtain $G'(f)$. It is obviously a radical change of epistemological position. This exchange generates the concept of *functional* and its symbolism $J(f)$ such as we saw in the calculus of variations above.

One then understands all the 'philosophical' enthusiasm of Comte for the "hypertranscendent" character of the calculus of variations.

CHAPTER XV
AFTER LEIBNIZ: SOME MODERN
EPISTEMOLOGICAL ASPECTS OF THE TRANSCENDENCE

ON THE CONCEPTS OF LEIBNIZ TODAY

The various obstacles encountered by the protagonists, and mentioned throughout this book, were not only linked to the complexity of the calculations. On the contrary, as we have seen, they mostly originated from the difficulty for the mathematicians of handling relevant conceptualisations of transcendence and algebraicity.

In this chapter, we intend to make a (very brief) overview of the continuation of the story; that is to say, the modern impact of the definitions by Leibniz of this pair of opposites. Firstly, we highlight this obvious fact; whether it is about numbers, about functions, or about curves, the distinction is made each time by characterizing the algebraic objects. The transcendent objects are simply the non-algebraic objects.

Algebraic functions, transcendent functions. An initial approach

In the first analysis, the presentation of the algebraic functions of *a single real variable*[1] can be done in a synthetic and ostensive mode. One can indeed produce a method of construction of the algebraic functions in the following way; both functions, $x \to 1$ and $x \to x$, are algebraic and we decide that, if f and g are algebraic functions, it will be the same of $f+g$, of $f-g$, and of $f.g$ (and thus of $f^2, f^3 \ldots, f^n$; n natural integer), but also of $\dfrac{f}{g}$, and also of $\sqrt[n]{f}$ if it exists. So, polynomials are algebraic, just like the rational fractions, and the radicals over these. All algebraic functions of one variable are then built in this way. This was the conception of Descartes, just like that of Leibniz. Euler then undertook to thematize and explain it accurately in the preface of the *Introductio* (see above XIII-A). As this background is very intuitive, thus we know how to write successive algebraic functions which are more and more complex. We will call this provisionally algebraic – (1) or algebraic-*standard;* this definition seems quite intuitive. Nevertheless, we see below how we have to modify it.

---

[1]   We restrict here to the case of the field of real numbers – this is the only case envisaged by Leibniz and mathematicians of the seventeenth century.

Accordingly, $x \to a^x, x \to x^{\sqrt{2}}, x \to Log\, x, x \to \sin x$, and also $x \to e^{x^2}$ are natural candidates to be transcendent functions. Obviously, it will be true only when we have proved it ...

So, the production *ad libitum* of such algebraic functions of a variable is easy and within the register of the synthesis. Obviously the proof of the transcendence of such a function remains difficult because it is inevitably indirect; it involves reducing an hypothesis to the absurd. This procedure is, from Leibniz's time, the consequence of what I called above "the absolute impossibility of inventory", a content which is negatively defined[2], (so, coextensive with some 'mystery'). In other words, there is no positive property, (i. e. direct), that characterizes the transcendent functions, i. e., no definition *for themselves*. Epistemologically speaking, the difficulty is obviously the same when demonstrating the irrationality and the transcendence: it is rooted in the lack of definition of the objects *per se*.

This situation, rather exceptional in mathematics, led, in irrationality and in transcendence, to inescapable methods of proof, which are specific to these two domains, and to which the mathematician is absolutely forced. As we saw before, (Chapter XIII), Waldschmidt does not fail to emphasise a close relationship in terms of the structure of proofs between the two epistemological organizations, irrationality and transcendence.

Algebraic curves, transcendent curves. Modern definitions

On this point, we first return to Descartes. By carefully following his prescriptions, the "geometric" curves (that is to say, those which are "algebraic" to Leibniz) are exactly those which possess an equation, like;

$$F(x, y) = a_0(y)\, x^n + a_1(y)\, x^{n-1} + \ldots + a_{n-1}(y)\, x + a_n(y) = 0$$

where, says Descartes, for each value of y, we **construct** the roots with respect to x of the equation above. The $a_k(y)$ are rational functions of a real variable, namely y. By multiplying by a suitable denominator, we may as well assume that these are polynomials, so that F is exactly a real polynomial in two variables (which would later become complex, by extension if necessary). The curves which are not algebraic are called "mechanical" by Descartes, and "transcendent" by Leibniz. As we have seen, Leibniz was greatly interested in the transcendent curves, thus defined.

In these conditions, are the algebraic curves exactly defined by algebraic equations? We can here observe a slight discrepancy between two concepts of algebraicity. It is clear that if we have y = G(x), where G is an algebraic function in one variable and, if we define F by F(x,y) = y − G(x), then F(x,y) = 0 is actually the equation of an algebraic curve in the previous sense. Conversely, however, if one has:

$$F(x, y) = 0 = a_0(y)\, x^n + a_1(y)\, x^{n-1} + \ldots + a_{n-1}(y)\, x + a_n(y) = 0 - \text{relation (A)}$$

---

2   About some of the difficulties encountered by Leibniz in the intervention of the negation see
     [Fichant 1998a], 'L'origine de la négation', for example pages 97–99.

where $a_k(y)$ are real polynomials with respect to y, it is not true that any solution $x \rightarrow y$ implicitly defined by (A) is an algebraic function of one variable in the sense of algebraic – (1) as previously defined. As we now know, there are indeed some solutions that may not be expressible by radicals – a fact that was ignored in the seventeenth century.

On the supremacy of the symbolism (again)

It is therefore preferable to unify the presentation and to say that a function of one variable $x \rightarrow y(x)$ is algebraic – (2) if, and only if, it can be defined *implicitly* by a relation such as that $F(x,y) = 0$, where F is a real polynomial with two variables. A curve defined by par $x \rightarrow y(x)$ is then said to be algebraic – (2) if and only if y is an algebraic function of a variable, in the sense (2). This last meaning will be final: today, one thus defines the algebraic curves and the transcendent curves.

Therefore, I will repeat here about the functions what I wrote of in Chapter XIII, about the numbers and the supremacy of symbolism. The definition-(2) of the algebraicity for the functions definitively triumphed over definition-(1) for the same symbolic reasons as the definition (for the numbers) of Lambert had triumphed over that of Euler: we can write symbolically and easily *any* polynomial equation with integer coefficients, but not *any* quantity written with sums, differences, products, quotients and radicals. Once again, one discovers two different symbolical positions of the mathematician.

On the ontology of the plane curves

This involved the curves 'in Descartes' style'; that is to say, defined by an algebraic equation, $F=0$. Yet, as we saw before in chapter V-B, Leibniz was extensively engaged with a different kind of curves, such as the cycloid – we say it is parametrically defined. What about a general concept of curve today? Before dealing with transcendent curves in the modern sense, one must indeed return to the difficulties regarding the concept of curve. Apparently so elementary, yet as we saw before (in Chapter V), it had so much interest for Leibniz; pertinently, this is noted by Engelsman[3]. How is this organized today?

These issues are taken into account in the following way; a plane curve may be given in two distinct ways – either implicitly, or parametrically – and this ontological duality organizes a dialectic.

---

3   Cf. *supra* Chapter. V-A), *Leibniz and the 'algebraical' curves.*

Curves implicitly defined

We say that a plane curve (C) is *implicitly defined* if it is the set of all of the points M(x, y) of the plan, such as $F(x,y) = 0$, where F is a function of two real variables, (that is to say, a mapping from an open subset U of IR^2 towards IR). Of course, the case of curves explicitly defined by $y = g(x)$ is a (very simple) special case.

We say that the curve (C) defined implicitly, is an *algebraic* curve if F is itself algebraic, (i. e. F is a polynomial in two variables), and *transcendent* otherwise. The relation $x^Y + y^X = a$ is a very simple example, cited by Tschirnhaus (see above XI-A) of the equation of a transcendent implicit curve.

We say that the curve (C) is of class C^k if F is itself of class C^k. If $k \geq 1$, we determine the tangent to the current point of an implicit curve by a guiding vector

$\vec{N} = \overrightarrow{grad\ F} = \left(\dfrac{\partial F}{\partial x}, \dfrac{\partial F}{\partial y} \right)$ of the normal (if it is not null). Recall that the normal is

the straight line passing through the relevant point of the curve and orthogonal to the tangent. This method, inaugurated by Huygens (Cf. XII-B7), thus allows us to place the tangent and the sub-tangent at any point of such a curve.

On Huygens' cubic curve

We shall take, for example, the cubic (H) of Huygens[4], (it is fully considered in Chapter XII), that is the curve as implicitly defined by the equation $x\,(x^2 + y^2) = a^2 y$. By setting:

$$F(x,y) = x\,(x^2 + y^2) - a^2 y, \text{ we have}$$

$$\vec{N} = (3x^2 + y^2, 2xy - a^2) \text{ for a normal vector to (H) at any point}[5].$$

$$\text{And thus } \vec{T} = (a^2 - 2xy, 3x^2 + y^2) \text{ for a tangent vector.}$$

On this issue, see *supra* in chapter XII-B, *The "normal vector" method.*

Curves defined parametrically

On the other hand, a plane curve (C) is said to be *parametrically defined* if there exists:
a) An open subset W of IR
b) Two functions ϕ_1 and ϕ_2 from $W \rightarrow \mathbb{R}$ such that a point M(x,y) of the plane belongs to (C) if and only if there exists $t \in W$ such that

$$x = \phi_1\,(t) \text{ and } y = \phi_2\,(t)$$

---

4   Cf. *supra* Chapter XII-B, *The Transcendence, bone of contention between Huygens and Leibniz (1690) A Mathematical Study.*
5   It is easily checked that the two coordinates of this vector cannot be simultaneously zero.

The set of such points M (x, y) is called the *support* of the curve (C) (it is noted supp(C)). A curve (C) defined parametrically, is said to be of class C^k if, and only if ϕ_1 and ϕ_2 are of class C^k.

If $k \geq 1$, the components of the tangent vector to the curve at the current point M(t) can be obtained by differentiating the parametric equations, so as to obtain the first (non null) vector (with $p \leq k$) $\left((\phi_1)^{(p)}, (\phi_2)^{(p)}\right)(t)$, if it exists.

The cycloid as an example

The cycloid with parametric equations:

$$x(t) = a(t - \sin t) \text{ and } y(t) = a(1 - \cos t), \text{ with } t \in \mathbb{R}$$

is a curve of class C^∞. The first derivative vector is:

$$x'(t) = a(1 - \cos t) \text{ and } y'(t) = a \sin t$$

Thus, it is nonzero at any point M(t), such that t is not an integer multiple of 2π, so it guides the tangent line at the point. The second derivative vector at a point M(t) is:

$$x''(t) = a \sin t \text{ and } y'' (t) = a \cos t$$

In an exceptional point M ($2k\ \pi$) (k is an integer)[6], this vector is thus not zero and has the value:

$$x''(2k\ \pi) = 0 \text{ and } y'' (2k\ \pi) = a$$

it thus guides the tangent at such a point.

On the dialectic of the duality

Between curves implicitly and parametrically defined, there are two reciprocal standard problems:

Problem 1) Given a curve (C) implicitly defined, of class C^k, find of it, if it is possible, a parameterization.

Problem 2) Given a curve (C) parametrically defined, of class C^k, find, if it is possible, an implicit equation of its support.

If $k \geq 1$, the implicit functions' theorem allows, most of the time, to assert in theory the local existence of solutions to these two problems. They are however local results (in the neighbourhood of a point) and also theoretical (i. e. implicit: we do not know, most of the time, how to be explicit).

---

6   Such a point is said to be *stationary* on the curve under consideration.

Algebraic numbers, transcendent numbers

Finally, there are the algebraic numbers. After Lambert, an *algebraic number* is simply a root of an algebraic equation with rational coefficients (or integers, which is the same): a real number a is thus algebraic, if and only if there exists $f \in Z[x]$ such that $f(a) = 0$. So, in these conditions, $\sqrt{2}, \sqrt[3]{2}+\sqrt[3]{3}$ etc., are algebraic numbers. This supplies a first support to intuition; however, this is insufficient because we now know, since Abel and Galois, that there are roots of algebraic equations with integer coefficients which may not be writable with radicals.

A real number that is not algebraic is called *transcendent*; for instance, e (Hermite) and π (Lindemann), etc. Any attempt to demonstrate the transcendence of a number comes up against the same problem as already mentioned for the functions; namely the necessity of an indirect demonstration.

Let us note *in fine* that the set of the algebraic numbers is countable; i.e., they may be "numbered"[7], unlike that of the transcendent numbers, which are thus "far more numerous" (see above, in chapter XIII-D, Liouville's considerations on the "very extensive" class of such numbers).

A couple of similar definitions, (algebraic – transcendent), is obviously valid in the area of complex numbers – this context, obviously ignored in the seventeenth century, is now naturally privileged. In this complex context, it should be emphasized, there is an important *structural property* of the algebraic numbers. Namely, *their set forms a (commutative) field*. As we know, this important result (established by Dedekind) was the source of all the important work on the fields (and subfields) of algebraic numbers in the first half of the twentieth century(See for instance [Ellison 1986]).

Leibniz, as the "founder of discursivity" of mathematical transcendence

It is thus Leibniz, and only he, who is at the origin of the creation of all the concepts of transcendence which we have just mentioned, and which organizes today a fully fledged mathematical discipline. Admittedly, he did not describe all the aspects and and did not succeed in his attempt at inventory – no one, in truth, has ever been able to. However, his relevant criticism of the limitations of the Cartesian mathematical thought and his own original reflections were the indisputable source used by mathematicians of the coming centuries to develop all the facets of the transcendence. Once again, as Dascal stresses pertinently, Leibniz has proven to be a "founder of discursivity"[8].

---

7   That is to say, we can put them in a one-to-one correspondence with the set IN of natural numbers.

8   [Dascal 2008a], 62: "[Leibniz] deserves to be acknowledged, as far as his contribution to rationality and dialectic is concerned, not as the author of this or that particular method, but rather as what Foucault calls a "fondateur de discursivité". By these he means "authors who are not only the authors of their works, of their books. They have produced something more: the possibility of formation of other texts ... they have established an infinite possibility of discourses".

REFERENCES

LEIBNIZ

GM: LEIBNIZ Gottfried Wilhelm, *Mathematische Schriften* (7 vol.) Gerhardt. Berlin. 1849–1863 Repr. *facsimile* Olms. Hildesheim. 1971–2004.

GP: LEIBNIZ Gottfried Wilhelm, *Philosophische Schriften* (7 vol.) Gerhardt. Berlin. 1875–199. Repr. *facsimile* Olms. Hildesheim. 1978.

GBM: LEIBNIZ Gottfried Wilhelm, *Der Briefwechsel von G.W. Leibniz mit Mathematikern*. Gerhardt. Berlin. 1899. Repr. Olms. Hildesheim. 1962

C: *Opuscules et fragments inédits, extraits des manuscrits de la bibliothèque de Hanovre*. L. Couturat ed. 1903. Repr. Olms, 1988.

A: Academy Edition of the Works of Leibniz. There are 8 series. The volumes are referenced by the letter A followed by the serial number (Roman numerals), followed by the volume number. The document concerned is identified by its number, preceded by the letter N. The series III, *Mathematischer, naturwissenschaftlicher und technischer Briefwechsel* (8 volumes published) and the series VII (7 volumes published) are of particular concern here. Some volumes are available on the Internet.

DESCARTES

[A.T] DESCARTES René, *Œuvres* (13 vol.). Adam-Tannery edition. Cerf. Paris. 1897–1913. Reedition of the first 11 volumes starting from 1964. Vrin. Paris (paperback edition starting from 1996). The texts are referenced A.T, followed by the number of the volume in Roman numerals and by that of the page in Arabic numerals. A (French) edition of the *Geometrie* of 1637 is printed in [A.T VI], 367–485.

[Smith & Latham] DESCARTES René. *The Geometry of René Descartes, With a Facsimile of the First Edition*. Transl. D.E Smith and M. L. Latham. New-York: Dover, 1954.

[G49] DESCARTES René, *Geometria*. Leyde, Jean Maire, 1649 (x + 118 pages) (2 vols.). It is the first Latin translation of the 1637 edition, by Franz Van Schooten. It is followed by 1°) F. de Beaune's *Notae Breves;* 2°) Schooten's *Commentary;* 3°) *Additamentum. Geometria a Renato Descartes anno 1637 gallice edita, nunc autem ... in linguam latinam versa.*

[Regulae] DESCARTES René, *Règles utiles et claires pour la Direction de l'Esprit et la Recherche de la Vérité*. Martinus Nijhoff. La Haye. 1977. We referenced [Regulae] this translation by J.-L. Marion of *Regulae ad Directionem Ingenii* [A.T X], 349–469, with mathematical notes of P. Costabel. We shall sometimes also indicate by l. the number of the line considered. Note the recent publication in paperback of two French translations of the *Regulae* – from Jacques Brunschwig. Livre de Poche. Paris. 1997 – from Joseph Sirven. Vrin. Paris. 2003.

English translation: R. Stoothoff, *Rules for The Direction of the Mind*, in *The Philosophical Writings of Descartes* (vol. I), Cambridge University Press. Cambridge. 1985, 9–78.

GENERAL REFERENCES

Babbage 1827] BABBAGE Charles, 'On the influence of signs in mathematical reasoning', *Transactions of Cambridge Philosophical Society*, 1827, vol. II.

[Baker 1975] BAKER Alan, *Transcendantal Number Theory*. Cambridge University Press. Cambridge.1976.

[Bélaval 1976] BÉLAVAL Yvon, *Études leibniziennes. De Leibniz à Hegel*, Paris. Gallimard. 1976.

[Bélaval 1960] BÉLAVAL Yvon, *Leibniz critique de Descartes*, Gallimard. Paris. 1960.

[Bernoulli Jn 1724] BERNOULLI Jean, 'Methodus commodo et naturalis reducendi Quadraturas transcendentes cujusvis gradus ad Longitudines Curvarum algebraicarum', *Acta Eruditorum*, August 1724, 356 = *Œuvres*, Bd. 2, 591.

[Bernoulli Jn 1697] BERNOULLI Jean, 'Principia Calculi Exponentialum, seu percurrentium', *Acta Eruditorum* 16, 125–133. March 1697 = *Œuvres* Bd I, 393–400.

[Blay 1995] BLAY Michel, *Les "Principia" de Newton*, PUF. Paris. 1995.

[Breger 2016] BREGER Herbert, *Kontinuum, Analysis, Informales – Beiträge zur Mathematik und Philosophie von Leibniz*, (Herausgegeben von W. Li), Springer. 2016.

[Breger 2016a] BREGER Herbert, 'Mathematics as the substructure of Leibniz's metaphysics', in [Breger 2016], 91–103.

[Breger 2016b] BREGER Herbert, 'Analysis as a feature of 17th century mathematics', in [Breger 2016], 159–173.

[Breger 2008] BREGER Herbert, Leibniz' calculation with compendia', in *Infinitesimal differences. Controversies between Leibniz and his contemporaries* (U. Goldenbaum and D. Jesseph eds.) De Gruyter. Berlin (etc.) 2008. 185–198.

[Breger 2008a] BREGER Herbert, 'The Art of Mathematical Rationality', in *Leibniz: What kind of Rationalist?* (M. Dascal ed.). Springer 2008, 141–152. (=[Breger 2016], 185–194).

[Breger 1992] BREGER Herbert, 'Le continu chez Leibniz', *Le labyrinthe du continu* (J.M Salanskis et H. Sinaceur eds.). Springer. 1992 (= [Breger 2016], 127–135).

[Breger 1988] BREGER Herbert, 'Symmetry in Leibnizian Physics', *The Leibniz Renaissance*. Olschki. Florence. 1988, 23–42. (= [Breger 2016], 13–27).

[Breger 1986] BREGER Herbert, 'Leibniz Einführung der Transzendenten', *Studia Leibnitiana*. Sonderheft 14,1986, 119–132.

[Browder 1976] BROWDER Felix (ed.), *Mathematical developments arising from HILBERT PROBLEMS*, Proceedings of Symposia in PURE MATHEMATICS, American Mathematical Society. Providence. Rhode Island. 1976.

[Cajori, 1928] CAJORI Florian, *A History of Mathematical Notations* (Vol. I and II). The Open Court Publishing Company. La Salle. Illinois. 1928. Repr. (one vol.). Dover. 1993.

[Cantor 1874] CANTOR Georg, 'Über eine Eigenschaft der Inbegriffes aller reellen algebraischen Zahlen' *Crelles' Journal*. 77 (1874), 252–262. French translation revised and corrected by the author = 'Sur une propriété du système de tous les nombres algébriques réels', *Acta mathematica* 2 (1883), 305–310.

[Cartan 1967] CARTAN Henri, *Calcul différentiel*. Hermann. Paris. 1967.

[Chareix 2010] CHAREIX Fabien, 'Geometrization or mathematization. Christiaan Huygen's critiques of infinitesimal analysis in his correspondence with Leibniz', *The practice of reason. Leibniz and his controversies*. (M. Dascal ed.), Benjamins. Amsterdam/Philadelphia. 2010, 33–49.

[Chevalier 1954] CHEVALIER Jacques, *PASCAL. Œuvres complètes*. Gallimard. Paris. 1954.

[Child 1920] CHILD James Mark, *The Early Mathematical Manuscripts of Leibniz*. The Open Court Publishing Company. Chicago-Londres. 1920.

[Comte 1830] COMTE Auguste, *Cours de philosophie positive*, vol. I. Rouen Frères Paris.1830.

[Costabel 1962] COSTABEL Pierre, 'Notes relatives à l'influence de Pascal sur Leibniz', *Revue d'histoire des sciences et de leurs applications*. 1962, Tome 15, n°3–4, 369–374.

[Couturat 1903] *Opuscules et fragments inédits* de Leibniz, ed. Louis COUTURAT. F. Alcan. Paris 1903. Reprinted Olms. 1985.

[Craig 1693] CRAIG John, *Tractatus mathematicus de figurarum curvilinearum quadraturis*, Londres. 1693.

[Craig 1685] CRAIG John, *Methodus figurarum lineis rectis et curvis comprehensarum quadraturas determinandi*, Londres. 1685.

[Dascal 2010] DASCAL Marcelo (ed.), *The practice of reason. Leibniz and his controversies*. Benjamins. Amsterdam/Philadelphia. 2010.

[Dascal 2008] DASCAL Marcelo (ed.), *Leibniz: What kind of Rationalist ?* Springer. 2008.

[Dascal 2008a] DASCAL Marcelo (ed.), 'Leibniz's Two-Pronged Dialectic', in [Dascal 2008], 37–72.

[Dascal 2008b] DASCAL Marcelo, G.W. Leibniz. *The Art of Controversies*. Springer. 2008.

[Dascal 1978] DASCAL Marcelo, *La sémiologie de Leibniz*. Aubier-Montaigne. Paris. 1978.

[De Buzon 1995] DE BUZON Frédéric, 'L'harmonie: métaphysique et phénoménalité', *Revue de métaphysique et de morale* I, 1995, pp. 95–120.

[De Cues 2010] (DE) CUES Nicolas, *De la docte ignorance* (1440). Translation, introduction and notes by Jean-Claude Lagarrigue. Cerf. Paris. 2010.

[De Gandt 1986] DE GANDT François, 'Le style mathématique des *Principia* de Newton', *Revue d'Histoire des Sciences*, 1986, 39–3, 195–222.

[De Morgan 1852] DE MORGAN Augustus, '*On the early history of infinitesimals in England*', Philosophical Magazine Series 4, Volume 4, Issue 26, 1852.

[Descotes 2008] DESCOTES Dominique, 'Constructions du triangle arithmétique de Pascal', *Mathématiciens français du XVIIème siècle. Descartes, Fermat, Pascal* (M. Serfati et D. Descotes eds). Presses Universitaires Blaise Pascal. Clermont Ferrand. 2008, 239–280.

[Dijksterhuis 1939] DIJKSTERHUIS E. J., 'James Gregory and Christiaan Huygens', in: *GREGORY, Tercentenary Memorial Volume*. Turnbull. London 1939, 478–486.

[Duchesneau 2008] DUCHESNEAU François, 'Rule of Continuity and Infinitesimals in Leibniz's Physics', *Infinitesimal Differences, Controversies between Leibniz and his Contemporaries* (U. Goldenbaum and D. Jesseph eds.). De Gruyter. Berlin. 2008, 235–253.

[Edwards 1987] EDWARDS A.W.F. *Pascal's Arithmetical Triangle*. Charles Griffin, 1987.

[Ellison 1986] ELLISON W. & F. 'La théorie des nombres algébriques', in *Abrégé d'histoire des Mathématiques* (J. Dieudonné ed.), Paris. Hermann. 1986. 173 201.

[Encyclopédie] *Encyclopédie Méthodique: Mathématiques* (3 vol.). Panckoucke. Paris. 1784–1789. Reprinted *facsimile*. ACL. Paris. 1987.

[Engelsman 1984] ENGELSMAN S. B., *Families of Curves and the Origins of Partial Differentiation*, North Holland Mathematical Studies. 1984.

[Euler 2013] EULER Leonhard, *Introductio In Analysin Infinitorum*. Translated and annotated by Ian Bruce (2 vol.).
= http://www.17centurymaths.com/contents/introductiontoanalysisvol1.htm

[Euler 2010] 'On Establishing a Relationship Among Three or More Quantities', Translation of [Euler 1785] by Geoffrey Smith St. Mark's School = http://eulerarchive.maa.org/docs/translations/E591trans.pdf

[Euler 1835] EULER Leonhard, *Introduction à l'Analyse infinitésimale. Avec des notes et éclaircissements* Traduction française de [Euler 1748] par J.B. Labey. Bachelier. Paris.1835. Reprinted *facsimile*. A.C.L. Paris 1988.

[Euler 1785] EULER Leonhard, 'De relatione inter ternas pluresque quantitates instituenda', *Opuscula analytica* 2 (1785), 91–101 (14 août 1775) = *Opera Omnia* serie 1, vol A. *Commentationes analyticae,* 136–146

[Euler 1748] EULER Leonhard, *Introductio in Analysin Infinitorum* (2 vol.). Bousquet. Lausanne. 1748.

[Euler 1744] EULER Leonhard, 'De Fractionibus continuis Dissertatio', *Acad. Sci. Petropolitanae* 9 (1737), 1744, 98–137 = *Opera Omnia* ser 1. vol 14. *Commentationes analyticae*, 188–215 (E71 indicis Enestroemiani). English Transl. 'An Essay on Continued Fractions', by Myra F. Wyman and Bostwick F. Wyman, *Math Systems Theory* 18 (1985), 295–328.

[Fichant 1998] FICHANT Michel, *Science et métaphysique dans Descartes et Leibniz*. Presses Universitaires de France. Paris. 1998.

[Fichant 1998a] FICHANT Michel, 'L'origine de la négation', in *Science et métaphysique dans Descartes et Leibniz*. Presses Universitaires de France. Paris. 1998, 85–120.

[Fichant 1969] FICHANT Michel et PÉCHEUX Michel, *Sur l'histoire des sciences*. Maspéro. Paris. 1969.

[Finster 1988] *Leibnizian Lexicon. A dual concordance to Leibniz's Philosophische Schriften*, Compiled by Finster R. Hunter G., MacRae R.F., Miles M., Seager W.E., Olms-Weidmann. 1988.

[Foulquié 1969] FOULQUIÉ Paul, *Dictionnaire de la langue philosophique*. P.U.F. Paris. (1st ed.)1969.

[Fraser 2005] FRASER Craig, 'Joseph Louis Lagrange, Théorie Des Fonctions Analytiques. First Edition (1797)', *Landmark Writings in Western Mathematics 1640–1940* (I. Grattan-Guinness ed.), Elsevier, 258–276.

[Galilée 1638] GALILEO Galilei, *Discorsi e Dimostrazioni matematiche intorno a due Nuove Scienze*, Traduction française de Maurice Clavelin: Galilée, *Discours concernant deux sciences nouvelles*. PUF. Paris.1995.

[Gel'fond 1934] GEL'FOND A.O, 'On Hilbert's seventh problem', Dokl. Akad. Nauk. SSSR 2, 1–6. Izv. Akad. Nauk. SSSR 7, 623–630.

[Gel'fond 1929] GEL'FOND A.O 'Sur les nombres transcendants', *C.R.A.S.* Série A 189, 1224–1226.

[Goldenbaum 2008] GOLDENBAUM Ursula and JESSEPH Douglas (eds.) *Infinitesimal Differences, Controversies between Leibniz and his Contemporaries.* De Gruyter. Berlin. 2008.

[Goldenbaum 2008a] GOLDENBAUM Ursula, 'Indivisibilia Vera – How Leibniz Came to Love Mathematics', in [Goldenbaum 2008], 53–94.

[Granger 1994] GRANGER Gilles Gaston, *Formes opérations, objets.* Vrin. Paris. 1994.

[Grattan-Guinness 2005] GRATTAN-GUINNESS Ivor (ed.) *Landmark Writings in Western Mathematics 1640–1940.* Elsevier. 2005.

[Grattan-Guinness 2004] GRATTAN-GUINNESS Ivor (ed.) *Companion Encyclopaedia Of The History And Philosophy Of The Mathematical Sciences*: 2 vol. Routledge. 2004 (repr. 2013)

[Grattan-Guinness 1997] GRATTAN-GUINNESS Ivor (ed.) The *Norton History of Mathematical Sciences*. Norton. New York-London. 1997.

[Gray 1978] GRAY, J.J. & TILLING L. 'Johann Heinrich Lambert, mathematician and scientist 1728–1777', *Historia mathematica*, Toronto, 1, 1978, 13–41.

[Gregory D. 1694] GREGORY David, *Memoranda*. July 1694. In *Correspondance de Newton*. Turnbull. Cambridge. 1961. Vol. III, N° 461, 385–387.

[Gregory J. 1667] GREGORY James, *Vera Circuli et Hyperbolae Quadratur*a. Padua.1667.

[Grosholz 2008] GROSHOLZ Emily, 'Productive Ambiguity in Leibniz's Representation of Infinitesimals', *Infinitesimal Differences, Controversies between Leibniz and his Contemporaries* (U. Goldenbaum & D. Jesseph eds.). De Gruyter. Berlin. 2008, 154–170.

[Grosholz 2007] GROSHOLZ Emily, *Representation and Productive Ambiguity in Mathematics and the Sciences*, Oxford University Press. 2007.

[Guicciardini 2005] GUICCIARDINI Niccoló, '1687. Isaac Newton, Philosophia naturalis principia mathematica', *Landmark Writings in Western Mathematics 1640–1940* (I. Grattan-Guinness ed.), Elsevier, 59–87.

[Hardy 1938] HARDY Godfrey Harold et WRIGHT Edward M. *Introduction to the Theory of Numbers*. Clarendon Press. Oxford. 1938 1st ed.; 2008 6th ed.

[Hemily 2006] H. E. M. I. L. Y – Collectif IREM de Lyon, *Textes fondateurs du calcul infinitésimal* (ed. O. Keller). Ellipses. Paris. 2006.

[Henry 1881] HENRY Charles, 'Étude sur le Triangle Harmonique', *Bulletin des Sciences Mathématiques et Astronomiques 2° série*, vol. 5 (N° 1) 1881, 96–113.

[Hermite-Stieltjes] *Correspondance Hermite-Stieltjes*, with a foreword by Émile Picard. Gauthier-Villars. Paris. Vol. 1, 8th November 1882 – 22nd July 1889 – Vol. 2, 18th October 1889 – 15th December 1894.

[Hermite 1873a] HERMITE Charles, 'Sur la fonction exponentielle', *C.R.A.S.* (1873) 18–24 & 74–79 & 226 – 233 & 285 – 293.

[Hermite 1873b] HERMITE Charles, 'Sur l'irrationalité de la base des logarithmes hyperboliques', *C.R.A.S.* (1873) = *Œuvres Complètes,* Published under the auspices of the Académie des Sciences. Gauthier-Villars. Paris. 1905, vol. III, 127–130.

[Hermite 1873c] HERMITE Charles, 'Sur quelques approximations algébriques'. Extract from a letter from Ch. Hermite to M. Borchardt. Crelle's Journal 76 (1873), 342–344.

[Hilbert 1900] *Mathematical Problems (lecture delivered before the International Congress of Mathematicians at Paris in 1900*. Proceedings of Symposia in pure Mathematics Volume 8

(1976), 1–34) = Göttinger Nachrichten 1900, 253–297. Translated by Laugel: 2nd International Congress of Mathematicians in Paris in 1900. Gauthier-Villars. Paris. 1902, 59–114.

[Hofmann 1976] HOFMANN Joseph, Introduction to volume III, I, of the Academy Edition of *Leibniz' Works* = Gottfried Wilhelm Leibniz. *Sämtliche Schriften und Briefe. Dritte Reihe: Mathematischer, naturwissenschaftlicher und technischer Briefwechsel.* Bd. 1: 1672–1676. Berlin. 1976, pp. I–LXXV.

[Hofmann 1974] HOFMANN Joseph, *Leibniz in Paris. 1672–1676. His Growth to Mathematical Maturity,* Cambridge University Press. 1974. English translation of *Die Entwicklungsgeschichte Mathematik während des Aufenthalts in Paris (1672–1676).* Oldenbourg. Munich. 1949.

{Hôpital 1696], (DE) L'HÔPITAL Guillaume, *Analyse des Infiniment Petits pour l'Intelligence des Lignes Courbes.* Paris. Imprimerie Royale.1696.

[Huygens 1888] HUYGENS Christiaan, *Œuvres complètes publiées par la Société Hollandaise des Sciences* (22 vol.). La Haye. 1888–1950.

[Huygens 1674] HUYGENS Christiaan, 'Le développement du 'Numerus impossibilis en série par Leibniz' .1674. *Œuvres complètes.* Vol. XX. *Musique et mathématique* (ed. J.A. Vollgraff). Martinus Nijhoff. Den Haag. 1940.

[Imbert 1992] IMBERT Claude, *Phénoménologies et langues formulaires.* Paris. PUF. 1992.

[Jesseph 2008], JESSEPH Douglas, 'Truth in Fiction: Origins and Consequences of Leibniz's Doctrine of Infinitesimal Magnitudes', *Infinitesimal Differences, Controversies between Leibniz and his Contemporaries* (U. Goldenbaum and D. Jesseph eds.). De Gruyter. Berlin. 2008, 215–233.

[Keller 2009] KELLER Olivier 'Le calcul différentiel de Leibniz appliqué à la chaînette',http://bibnum.education.fr/mathématiques/analyse/le-calcul-différentiel-de-leibniz-appliqué-à-la-chaînette.

[Kuzmin 1830] KUZMIN R.O, 'On a new class of transcendental numbers', Izv. Akad. Nauk. SSSR, Ser. Math. 3, 585–597.

[Lalande 1927] LALANDE André, *Vocabulaire technique et critique de la Philosophie.* Paris. Alcan. 1927. Rep. Presses Universitaires de France, Paris. 2006.

[Lambert 1770] LAMBERT Jean-Henri, 'Observations trigonométriques', Mémoire de l'Académie des Sciences de Berlin [24] (1768), 1770, 327–354 = Œuvres, Vol. 2, 245–269.

[Lambert 1761] LAMBERT Jean-Henri, 'Mémoire sur quelques propriétés remarquables des quantités transcendantes circulaires et logarithmiques', *Mémoires de l'Académie des Sciences de Berlin* [17], (1761), 1768, 265–322 = Œuvres, 2, Zurich, A. Speiser, 1946, 112–159. = http://www.bibnum.education.fr/mathématiques/théorie-des-nombres/lambert-et-l'irrationalité-de-π–1761

[Lebesgue 1950] LEBESGUE Henri, *Leçons sur les Constructions géométriques.* Paris. Gauthier Villars. 1950. Reprinted Gabay. Paris 1987.

[Lebesgue 1932] LEBESGUE Henri, 'Sur les démonstrations dites élémentaires de la transcendance de *e* et de π', *L'enseignement Scientifique.* Genève. June 1932, 217–267.

[Legendre 1842] LEGENDRE Adrien Marie, *Éléments de Géométrie.* Firmin Didot. Paris. Edition 1842.

[Lindemann 1882] LINDEMANN C.F, ‚Über die Zahl π', *Mathematische Annalen* 20 (1882), 213–225.

[Liouville 1844a] LIOUVILLE Joseph, 'Sur des classes très-étendues de quantités qui ne sont ni algébriques, ni même réductibles à des irrationnelles algébriques', *C.R.A.S.* 18(1844), 883–885.

[Liouville 1844b] LIOUVILLE Joseph, 'Sur des classes très-étendues de quantités qui ne sont ni algébriques, ni même réductibles à des irrationnelles algébriques', *C.R.A.S.* 18(1844), 910–911.

[Liouville 1844c] LIOUVILLE Joseph, 'Sur des classes très-étendues de quantités qui ne sont ni algébriques, ni même réductibles à des irrationnelles algébriques', *J.M.P.A* (1)16 (1851), 133–142.

[Mahnke 1926] MAHNKE Dietrich, 'Leibniz als Begründer der symbolischen Mathematik', *ISIS*, 30, IX/2 (1926), 279–293.

[Martineau 1853] MARTINEAU Harriet, *The Positive Philosophy of August Comte*, English translation & condensation of Auguste Comte's *Cours de philosophie positive*. London. John Chapman. 1853.

[Massa Esteve 2009] MASSA ESTEVE M.R & DELSHAMS A, 'Euler beta integral in Pietro Mengoli's works', *Arch. Hist. Exact. Sci.* 63 (2009), 325–356.

[Mengoli 1672] MENGOLI Pietro, *Circolo*, Bologna. 1672.

[Mercator 1668] MERCATOR Nicolas, *Logarithmotechnia; sive methodus construendi logarithmos nova, accurata et facilis; scripto antehàc communicata, anno sc. 1667 nonis Augusti: cui nunc accedit vera quadratura hyperbolae, & inventio summae logarithmorum*. London. 1668.

[Mesnard 1964] MESNARD Jean, *Œuvres Complètes de Blaise Pascal* (4 vol.) Desclées de Brouwer. Paris.1964–1992.

[Milhaud 1921] MILHAUD Gaston. *Descartes savant*. Félix Alcan. Paris. 1921.

[Montucla 1799, I à IV] MONTUCLA Jean-Étienne, *Histoire des Mathématiques*. (4 vol.). Jombert. Paris. 1758 (1st edition). Agasse. Paris. 1799–1802 (2nd expanded edition). Reprint Blanchard. Paris. 1960.

[Montucla 1754] MONTUCLA Jean-Étienne, *Histoire des recherches sur la Quadrature du Cercle*. Jombert. Paris. 1754. Reprint I.R.E.M, University Paris VII. 1986.

[Newton 1759] NEWTON Isaac, *Principes mathématiques de la philosophie naturelle*. French translation of the *Principia* by Émilie du Chatelet. Preface by Voltaire. Reprint *facsimile*. Dunod. Paris. 2006.

[Newton 1732] NEWTON Isaac, *Arithmetica Universalis sive de Compositione et resolutione arithmetica liber*. Verbeek. Leyde. 1732 – Third edition, published by W.J.S van s'Gravesande and supplemented with texts of Halley, Moivre, MacLaurin, Campbell, etc. One of the Latin posthumous editions. Some others were published from the *princeps* edition (Cambridge 1707), and from a second one (London, 1722).

[Newton 1707] NEWTON Isaac, 'Aequationum Constructio linearis', in *Arithmetica Universalis*, page 212 in 1732 edition.

[Newton 1687] NEWTON Isaac, *Philosophia Naturalis Principia Mathematica*. Londres. 1687. English modern translation, by I. Bernard Cohen and Anne Whitman. University of California Press.1999.

[Pappus 1982] PAPPUS D'ALEXANDRIE, *La Collection Mathématique*, translation from Greek by Paul Ver Eecke, Paris, 1933, 2 vol. Repr. A. Blanchard. Paris.1982.

[Parmentier 2004] *Quadrature arithmétique du cercle, de l'ellipse et de l'hyperbole et la trigonométrie sans tables trigonométriques qui en est le corollaire*, French translation by M. Parmentier of *De quadratura arithmetica circuli ellipseos et hyperbolae cujus corollarium est trigonometria sine tabulis*. Latin text by E. Knobloch. Paris Vrin. 2004.

[Parmentier 1989] PARMENTIER Marc, *Leibniz: Naissance du calcul différentiel*. Vrin. Paris. 1989.

[Peiffer 1998] PEIFFER Jeanne, 'Faire des mathématiques par lettres', *Revue d'Histoire des Mathématiques* 4. 1998, 143–157.

[Peiffer 1989] PEIFFER Jeanne, 'Leibniz, Newton et leurs disciples', *Revue d'histoire des sciences*.1989, 42–3, 303–312.

[Pensivy 1986] PENSIVY Michel, *Jalons historiques pour une épistémologie de la série infinie du binôme*. Doctoral dissertation, Université de Nantes, 1986. Published in *Sciences et techniques en perspectives*. Université de Nantes. 14, 1987–1988.

[Philonenko 1967] PHILONENKO Alexis, 'La loi de continuité et le principe des indiscernables', *Revue de Métaphysique et de Morale* 72/1, 1967, 261–286.

[Plutarque 1877] PLUTARQUE, *Vie de Marcellus,* in *La Vie des Hommes Illustres de Rome*. Limoges. Eugène Ardant. 1877, page 22

[Pesic 2001] PESIC Peter, 'The validity of Newton's Lemma 28', *Historia Mathematica*, 28, 215–219.

[Probst 2012] PROBST Siegmund and MAYER Üwe, Introduction to vol. A VII, 6 of the Academy Edition (*Écrits mathématiques*, 1673–1676, *Quadrature arithmétique du cercle*).

[Probst 2008] PROBST Siegmund, 'Indivisible and Infinitesimals in Early Mathematical Texts of Leibniz', in *Infinitesimal differences. Controversies between Leibniz and his contemporaries* (U. Goldenbaum and D. Jesseph eds.) De Gruyter. Berlin (etc.) 2008. 95–106.

[Probst 2008a] PROBST Siegmund and MAYER Üwe, Introduction to vol. A VII, 5 of the Academy Edition (*Écrits mathématiques,* 1674–1675, *Mathématiques infinitésimales*).

[Probst 2007] PROBST Siegmund,'The Reception of Pietro Mengoli's Work on Series by Leibniz (1672–1676)', *Proceed. Joint Intern. Meeting UMI-DMV,* Perugia, 18–22 June 2007.

[Probst 2006] PROBST Siegmund, ,Differenzen, Folgen und Reihen bei Leibniz (1672–1676)', *Wanderschaft in der Mathematik* (M. Hyksova & U. Reich eds.), Rauner. Augsburg. 2006.

[Riesenfeld 2014] RIESENFELD Dana and SCARAFILE Giovanni (eds), *Perspectives on Theory of Controversies and the Ethics of Communication. Explorations of Marcelo Dascal's Contributions to Philosophy.* Springer. 2014.

[Roero 2005], ROERO Clara Sylvia, 'Gottfried Wilhelm Leibniz, First Three Papers on the Calculus (1684, 1686, 1693)', *Landmark Writings in Western Mathematics 1640–1940* (I. Grattan-Guinness ed.), Elsevier, 46–58.

[Schneider 1934], SCHNEIDER Th., 'Transzendenzuntersuchungen periodischer Funktionen. I: Transzendenz von Potenzen. II: Transzendenzeigenschaften elliptischer Funktionen'. *J. Reine Angew. Math.*, 172, 65–69, 70–74.

[Serfati 2016a], SERFATI Michel, 'Leibniz contre Descartes: l'invention de la transcendance mathématique. Les aventures d'une exceptionnelle création', *Actes Xème Internationalen Leibniz-Kongresses* (Wenchao Li ed.) Hannover. 18–23 July 2016. Olms. 2016, 211–222.

[Serfati 2016b], SERFATI Michel, 'Leibniz y la invención de la trascendencia matemática. Las peripecias de un imposible inventorio', *La monadología de Leibniz a debate* (J.A. Nicolas, M. Sánchez, M. Escribano, L. Herrera, M. Higueras, M. Palomo, J.M. Gómez Delgado eds). Editorial Comares. 2016, 25–36.

[Serfati 2016c], SERFATI Michel, 'Harmonie, symbolisme et structures dans les mathématiques de Leibniz', *O labirinto da harmonia. Estudos sobre Leibniz*, Bibliothèque Nationale du Portugal. Catalogue de Publications. Lisbonne. Novembre 2016, 59–84.

[Serfati 2013a], SERFATI Michel, 'Order in Descartes, Harmony in Leibniz: Two Regulative Principles of Mathematical Analysis'. *Studia Leibnitiana*. Band 45 (2013) Heft 1, 59–96.

[Serfati 2013b] SERFATI Michel, 'On the sum of all differences' and the origin of mathematics, according to Leibniz: Mathematical and philosophical aspects', in *Perspectives on Theory of Controversies and the Ethics of Communication, Writings in honor of Marcelo Dascal* (D. Riesenfeld and G. Scarafile eds.), Springer 2013, 69–80.

[Serfati 2012], SERFATI Michel, Review of volume A III, 7 of Leibniz' Mathematical Works in the Academy Edition. Steiner. Stuttgart. Band XLII. Heft 2, 244–256.

[Serfati 2011a] SERFATI Michel, 'Mathematical and philosophical aspects of the Harmonic Triangle in Leibniz', *Proceed. IX Internat. Leibniz Kongress* (H. Breger, J. Herbst, S. Erdner eds). Vol. III, 1060–1069. Hannover. Sept. 2011.

[Serfati 2011b] SERFATI Michel, 'Le développement de la pensée mathématique du jeune Descartes, in *De la Méthode* (M. Serfati ed.). Presses Universitaires de Franche Comté. Besançon. 2011 (2nd ed.), 43–77.

[Serfati 2011c] SERFATI Michel, 'Analogies et "prolongements". Permanence des formes symboliques et constitution d'objets mathématiques', in *De la Méthode*. (M. Serfati ed.). Presses Universitaires de Franche Comté. Besançon. 2011 (2nd ed.), 265–313.

[Serfati 2011d], SERFATI Michel, 'Dascal, Leibniz, et le symbolisme mathématique', *A Crua Palavra* (G. Scarafile ed.), pp. I-XIII.

[Serfati 2010a] SERFATI Michel, 'The principle of continuity and the 'paradox' of Leibnizian mathematics', *The Practice of Reason: Leibniz and his Controversies*. (M. Dascal ed.) Benjamins (Controversies, volume 7). Amsterdam. 2010. 1–32.

[Serfati 2010b], SERFATI Michel, 'Symbolic revolution, scientific revolution: mathematical and philosophical aspects', *Philosophical Aspects of Symbolic Reasoning in Early Modern Mathematics* (A. Heeffer and M. Van Dyck eds.), Studies in Logic 26, London. College Publications, 2010, pp. 105–124.

[Serfati 2008a] SERFATI Michel, 'Symbolic inventiveness and "irrationalist" practices in Leibniz' mathematics', in *Leibniz: What kind of rationalist* ? (M. Dascal ed.). Springer. 2008. 125–139.

[Serfati 2008b] SERFATI Michel, 'Constructivismes et obscurités dans la *Géométrie* de Descartes', *Mathématiciens français du XVIIe siècle: Pascal, Descartes, Fermat* (M. Serfati and D. Descotes eds.), Clermont-Ferrand, Presses Universitaires Blaise Pascal, 2008, 11–44.

[Serfati 2008c] SERFATI Michel, 'L'avènement de l'écriture symbolique mathématique. Symbolisme et création d'objets', *Lettre de l'ACADEMIE des SCIENCES* 24 (automne 2008).

[Serfati 2006a] SERFATI Michel, 'Leibniz's Practice of Harmony in Mathematics', Act. Coll. *Leibniz, Einheit in der Vielheit* (H. Breger, J. Herbst, S. Erdner eds.), Hannover July 24–29 2006, Vol. 2, 974–981.

[Serfati 2006b] SERFATI Michel, 'La constitution de l'écriture symbolique mathématique. Symbolique et invention', *Gazette des Mathématiciens* 108 (April 2006), Soc. Math. Fr., 101–118.

[Serfati 2005] SERFATI Michel, *La révolution symbolique. La constitution de l'écriture symbolique mathématique*. Paris. Pétra. 2005.

[Serfati 2005a] SERFATI Michel, '1649. René Descartes, Geometria', *Landmark Writings in Western Mathematics 1640–1940* (I. Grattan-Guinness ed.), Elsevier, 1–22.

[Serfati 2001] SERFATI Michel, 'Mathématiques et pensée symbolique chez Leibniz', *Mathématiques et physique leibniziennes* (1ère partie) (M. Blay et M. Serfati dirs.), *Revue d'Histoire des Sciences*, 54–2 (2001), 165–222.

[Serfati 1999] SERFATI Michel, 'La dialectique de l'indéterminé, de Viète à Frege et Russell', *La recherche de la vérité* (M. Serfati ed.). A.C.L. Paris. 1999, 145–174.

[Serfati 1998] SERFATI Michel, 'Descartes et la constitution de l'écriture symbolique mathématique', *Pour Descartes* (M. Serfati éd.), *Revue d'Histoire des Sciences* 51–2/3 (1998), 237–290.

[Serfati 1995] SERFATI Michel, *Exercices de Mathématiques pour Maths Sup. et Spé. – Géométrie & Cinématique,* Belin. Paris. 1995.

[Serfati 1995a] SERFATI Michel, *Exercices de Mathématiques pour Maths Sup. et Spé. – Analyse II,* Belin. Paris. 1995.

[Serfati 1994] SERFATI Michel, '*Regulae* et Mathématiques', *Theoria* (San Sebastián), *Segunda Epoca* IX, n° 21 (1994), 61–108.

[Serfati 1993] SERFATI Michel, 'Les compas cartésiens', *Archives de Philosophie* 56 (1993), 197–230.

[Serfati 1992] SERFATI Michel, *Quadrature du cercle, fractions continues, et autres contes*. Éditions de l'Association des Professeurs de Mathématiques. Paris. 1992.

[Serfati 1967] SERFATI Michel, 'Problème 11', *Problèmes de mathématiques*, SEDES. Paris 1967, 123–129.

[Stewart 1989] STEWART Ian, *Galois Theory*. Chapman and Hall. Londres/New York. 1989.

[Sturm 1689] STURM John, *Mathesis enucleata, Cujus Praecipua Contenta Sub Finem Praefationis, uno quasi obtutu spectanda exhibentur*. Nuremberg. 1689.

[Tidjeman 1976] TIDJEMAN Robert, 'On the Gel'fond-Baker method and its applications', *Proccedings of Symposia in Pure Mathematics*, AMS Publications. Volume 28, 1976, 241–268.

[Turriere 1919] TURRIERE Émile, 'La notion de transcendance géométrique chez Descartes et Leibniz', *Isis* 2 (1919), 106–124.

[Viète 1591] VIÈTE François, *In artem analyticem Isagoge sursim excussa ex opere restitutae mathematicae analyseos seu Algebra nova*. Mettayer. Tours. 1591. French translation J.-L. de Vaulézard (Jacquin. Paris. 1630), *Introduction en l'art analytic ou nouvelle algèbre de François Viète*. Reprint Corpus des œuvres de philosophie en langue française. Fayard. Paris. 1986.

[Vilain 2009] VILAIN Christiane, *Naissance de la physique moderne – Méthode et philosophie mécanique au XVIIe siècle*. Ellipses. Paris. 2009.

[Vuillemin 1962] VUILLEMIN Jules, *La Philosophie de l'Algèbre* (vol. 1). Presses Universitaires de France. Paris. 1962.

[Wahl 2011] WAHL Charlotte, Assessing Mathematical Progress. Contemporary Views on the Merits of Leibniz's Infinitesimal Calculus', *Proceed. IX Internat. Leibniz Kongress* (H. Breger, J. Herbst, S. Erdner eds). Vol. III, 1174–1182. Hannover. Sept. 2011.

[Waldschmidt 2012] WALDSCHMIDT Michel, 'Questions de transcendance: grandes conjectures, petits progrès', Journées annuelles de la Société Mathématique de France. Paris 2012, pp. 49 to 67 http://www.math.jussieu.fr/~miw/articles/pdf/TranscendanceSMF2012.pdf

[Waldschmidt 2003] WALDSCHMIDT Michel, 'Quelles sont les méthodes permettant de montrer qu'un nombre est transcendant?' Proc. *Journées "Sur les méthodes en mathématiques"*, Paris. April 2003.

[Waldschmidt 2000] WALDSCHMIDT Michel, *Diophantine Approximation on Linear Algebraic Groups. Transcendence Properties of the Exponential Function in Several Variables*, Grundlehren der Mathematischen Wissenschaften 326, Springer-Verlag, Berlin-Heidelberg. 2000.

[Waldschmidt 1999] WALDSCHMIDT Michel, 'Les débuts de la théorie des nombres transcendants', *Cahiers du Séminaire d'Histoire des Mathématiques*, Publ. Univ. P. et M. Curie. (Lab. Math. Fond. – E.H.E.S.S) 4 (1983), 93–115 = *La recherche de la vérité* (M. Serfati ed.). Paris. A. C. L. 1999, 73–96.

[Wantzel 1837], WANTZEL Pierre Laurent, 'Recherche sur les moyens de reconnaître si un problème de géométrie peut être résolu par la règle et le compas', J.M.P.A (1837), 366–372.

[Whiteside 1961] WHITESIDE Derek Thomas, 'Patterns of mathematical thought in the later seventeenth century', *Archive for History of Exact Sciences* 1 (1961), 174–388.

[Youschevitch 1976] YOUSCHEVITCH A.P, 'The concept of function up to the mIddle of the 19th century', *Archive for History Of Exact Sciences* 16/1 (1976–1977), 37–85.

STUDIA LEIBNITIANA — SONDERHEFTE

Im Auftrag der Gottfried-Wilhelm-Leibniz-Gesellschaft e.V.
herausgegeben von Herbert Breger, Wenchao Li, Heinrich Schepers und Wilhelm Totok †.

Franz Steiner Verlag   ISSN 0341-0765

2.   George Henry R. Parkinson
     Leibniz on Human Freedom
     1970. VI, 67 S., kt.
     ISBN 978-3-515-00272-1

3.   Kurt Müller / Heinrich Schepers /
     Wilhelm Totok (Hg.)
     Linguistik und Sprachstudium
     Symposion der Leibniz-Gesellschaft
     Hannover vom 15.–16. November 1971
     1973. VIII, 174 S., kt.
     ISBN 978-3-515-00273-8

4.   Kenneth C. Clatterbaugh
     **Leibniz's Doctrine of Individual
     Accidents**
     1973. VIII, 92 S., kt.
     ISBN 978-3-515-00274-5

5.   **Der Wissenschaftsbegriff
     in der Natur- und in den Geistes-
     wissenschaften**
     Symposion der Leibniz-Gesellschaft
     Hannover vom 23.–24. November 1973
     1975. VIII, 302 S. mit 3 Abb., 1 Tab. und
     2 Schemata, kt.
     ISBN 978-3-515-02109-8

6.   **Die Bedeutung der Wissenschafts-
     geschichte für die Wissenschafts-
     theorie**
     Symposion der Leibniz-Gesellschaft
     Hannover vom 29.–30. November 1974
     1977. VIII, 170 S., kt.
     ISBN 978-3-515-02394-8

7.   **Magia Naturalis und die Entstehung
     der modernen Naturwissenschaften**
     Symposion der Leibniz-Gesellschaft
     Hannover vom 14.–15. November 1975
     1978. VIII, 180 S., kt.
     ISBN 978-3-515-02778-6

8.   Albert Heinekamp / Franz Schupp (Hg.)
     **Die intensionale Logik bei Leibniz
     und in der Gegenwart**
     Symposion der Leibniz-Gesellschaft
     Hannover vom 10.–11. November 1978
     1979. IX, 153 S., kt.
     ISBN 978-3-515-03011-3

9.   George Henry R. Parkinson (Hg.)
     Truth, Knowledge and Reality
     Inquiries into the Foundations of
     Seventeenth Century Rationalism.
     A Symposium of the Leibniz-Gesellschaft
     Reading, 27th–30th July 1979
     1981. IX, 158 S., kt.
     ISBN 978-3-515-03350-3

10. Albert Heinekamp (Hg.)
     Leibniz als Geschichtsforscher
     Symposion des Istituto di Filosofici Enrico
     Castelli und der Leibniz-Gesellschaft
     in Ferrara vom 12.–15. Juni 1980
     1982. XI, 186 S. mit 6 Abb., kt.
     ISBN 978-3-515-03647-4

11. Diogenes Allen
     **Mechanical Explanations and
     the Ultimate Origin of the Universe
     According to Leibniz**
     1983. V, 44 S., kt.
     ISBN 978-3-515-03867-6

12. Werner Kutschmann
     Die Newtonsche Kraft
     Metamorphose eines wissenschaftlichen
     Begriffs
     1983. VIII, 177 S., kt.
     ISBN 978-3-515-03727-3

13. Albert Heinekamp (Hg.)
     Leibniz' Dynamica
     Symposion der Leibniz-Gesellschaft
     in der Evangelischen Akademie Loccum
     vom 2.–4. Juli 1982
     1984. 226 S. mit 5 Abb., kt.
     ISBN 978-3-515-03869-0

14. Albert Heinekamp (Hg.)
     **300 Jahre „Nova Methodus"
     von G. W. Leibniz (1684–1984)**
     Symposion der Leibniz-Gesellschaft
     im Congresscentrum „Leewenhorst"
     in Nordwijkerhout (Niederlande)
     vom 28.–30. August 1984
     1987. XVI, 268 S., kt.
     ISBN 978-3-515-04470-7

15. Albert Heinekamp (Hg.)
     Leibniz: Questions de logique
     Symposion organisé par la Gottfried-
     Wilhelm-Leibniz-Gesellschaft e.V.
     Hannover, Bruxelles, Louvain-la-Neuve,

26–28 Août 1985
1988. XIV, 208 S., kt.
ISBN 978-3-515-04604-6

16. Hans Poser / Albert Heinekamp (Hg.)
Leibniz in Berlin
Symposion der Leibniz-Gesellschaft
und des Instituts für Philosophie, Wissen-
schaftstheorie, Wissenschafts- und Tech-
nikgeschichte der Technischen Universität
Berlin vom 10.–12. Juni 1987
1990. 305 S., kt.
ISBN 978-3-515-05056-2

17. Heinz-Jürgen Heß / Fritz Nagel (Hg.)
**Der Ausbau des *Calculus* durch
Leibniz und die Brüder Bernoulli**
1989. 175 S. mit 62 Abb., kt.
ISBN 978-3-515-05082-1

18. Claudia von Collani (Hg.)
**Vorschlag einer päpstlichen
Akademie für China**
Ein Brief des Chinamissionars Joachim
Bouvet an Gottfried Wilhelm Leibniz und
an den Präsidenten der Académie des Sci-
ences Jean-Paul Bignon aus dem Jahre
1704
1989. 136 S., kt.
ISBN 978-3-515-05186-6

19. Helmut Pulte
**Das Prinzip der kleinsten Wirkung
und die Kraftkonzeption der
rationalen Mechanik**
1990. XI, 309 S., kt.
ISBN 978-3-515-04984-9

20. Erhard Holze
**Gott als Grund der Welt im Denken
des Gottfried Wilhem Leibniz**
1991. 204 S., kt.
ISBN 978-3-515-05803-2

21. Gottfried Wilhelm Leibniz
Le Meilleur des Mondes
Hg. von Albert Heinekamp und André
Robinet
1992. 295 S., kt.
ISBN 978-3-515-05764-6

22. Renato Cristin (Hg.)
**Leibniz und die Frage nach
der Subjektivität**
Leibniz-Tagung in Triest vom 11.–14. Mai
1992
1994. 229 S., kt.
ISBN 978-3-515-06230-5

23. Susanne Edel
**Metaphysik Leibnizens
und Theosophie Böhmes**
Die Kabbala als *Tertium Comparationis*
für eine rezeptionsgeschichtliche Unter-
suchung der individuellen Substanz
1995. 225 S., kt.
ISBN 978-3-515-06666-2

24. Martine de Gaudemar (Hg.)
La notion de nature chez Leibniz
Colloque organisé par le departement de
philosophie de l'université de Provence
(Aix-en-Provence), le CNRS (Paris), et la
G.W. Leibniz-Gesellschaft (Hannover),
Aix-en-Provence, 13–15 Octobre 1993
1995. 240 S., kt.
ISBN 978-3-515-06631-0

25. Alexander Wiehart-Howaldt
Essenz, Perfektion, Existenz
Zur Rationalität und dem systematischen
Ort der Leibnizschen *Theologia Naturalis*
1996. XII, 223 S., kt.
ISBN 978-3-515-06840-6

26. Emily Grosholz / Elhanan Yakira
Leibniz's Science of the Rational
1998. 107 S., kt.
ISBN 978-3-515-07400-1

27. Paul Blum
**Philosophenphilosophie
und Schulphilosophie**
Typen des Philosophierens in der Neuzeit
1998. 302 S., kt.
ISBN 978-3-515-07201-4

28. Herbert Breger /
Friedrich Niewöhner (Hg.)
Leibniz und Niedersachsen
Tagung anläßlich des 350. Geburtstages
von G. W. Leibniz, Wolfenbüttel 1996
1999. 238 S. und 16 Farbtaf., kt.
ISBN 978-3-515-07200-7

29. Martin Fontius / Hartmut Rudolph /
Gary Smith (Hg.)
Labora diligenter
Potsdamer Arbeitstagung zur Leibniz-
forschung vom 4.–6. Juli 1996
1999. 240 S., kt.
ISBN 978-3-515-07602-9

30. Brandon Look
**Leibniz and the
'Vinculum Substantiale'**
1999. 143 S., kt.
ISBN 978-3-515-07623-4

31. Andreas Hüttemann (Hg.)
**Kausalität und Naturgesetz
in der Frühen Neuzeit**
2001. 240 S., kt.
ISBN 978-3-515-07858-0

32. Massimiliano Carrara / Antonio-Maria
Nunziante / Gabriele Tomasi (Hg.)

Individuals, Minds and Bodies
Themes from Leibniz
2004. 297 S., kt.
ISBN 978-3-515-08342-3

33. Alexandra Lewendoski (Hg.)
Leibnizbilder
im 18. und 19. Jahrhundert
2004. 261 S., kt.
ISBN 978-3-515-08401-7

34. Daniel J. Cook / Hartmut Rudolph /
Christoph Schulte (Hg.)
Leibniz und das Judentum
2008. 283 S. mit 6 fbg. und 1 s/w-Abb., kt.
ISBN 978-3-515-09251-7

35. Mark Kulstad / Mogens Lærke /
David Snyder (Hg.)
The Philosophy of the Young Leibniz
2009. 259 S. mit 1 Abb., kt.
ISBN 978-3-515-08098-9

36. Paul Rateau (Hg.)
L'idée de théodicée de Leibniz
à Kant: héritage, transformations,
critiques
2009. 222 S., kt.
ISBN 978-3-515-09351-4

37. Juan Antonio Nicolás (Hg.)
Leibniz und die Entstehung
der Modernität
Leibniz-Tagung in Granada,
1.–3. November 2007
2010. 278 S., kt.
ISBN 978-3-515-09357-6

38. Erich Barke / Rolf Wernstedt /
Herbert Breger (Hg.)
Leibniz neu denken
2009. 108 S., kt.
ISBN 978-3-515-09374-3

39. Thomas Kisser (Hg.)
Metaphysik und Methode
Descartes, Spinoza, Leibniz im Vergleich
2010. 153 S., kt.
ISBN 978-3-515-09736-9

40. Paul Rateau (Hg.)
Lectures et interprétations des
Essais de théodicée de G. W. Leibniz
2011. 316 S. mit 2 Abb., kt.
ISBN 978-3-515-09919-6

41. Wenchao Li / Hans Poser /
Hartmut Rudolph (Hg.)
Leibniz und die Ökumene
2013. 314 S., kt.
ISBN 978-3-515-10309-1

42. Wenchao Li / Hartmut Rudolph (Hg.)
„Leibniz" in der Zeit
des Nationalsozialismus
2013. 309 S., kt.

ISBN 978-3-515-10308-4

43. Christian Leduc / Paul Rateau /
Jean-Luc Solère (Hg.)
Leibniz et Bayle :
Confrontation et dialogue
2015. 452 S., kt.
ISBN 978-3-515-10638-2

44. Wenchao Li (Hg.)
„Das Recht kann nicht
ungerecht sein ..."
Beiträge zu Leibniz' Philosophie
der Gerechtigkeit
2015. 184 S. mit 3 Abb., kt.
ISBN 978-3-515-11212-3

45. Arnaud Pelletier (Hg.)
Leibniz and the aspects of reality
2016. 149 S. mit 7 Abb., kt.
ISBN 978-3-515-11170-6

46. Arnaud Pelletier (Hg.)
Leibniz's experimental philosophy
2016. 257 S. mit 16 Abb., kt.
ISBN 978-3-515-11307-6

47. Wenchao Li / Simona Noreik (Hg.)
Leibniz, Caroline und die Folgen
der englischen Sukzession
2016. 136 S. mit 8 fbg. und 2 s/w-Abb., kt.
ISBN 978-3-515-11383-0

48. Ansgar Lyssy
Kausalität und Teleologie
bei G. W. Leibniz
2016. 417 S., kt.
ISBN 978-3-515-11349-6

49. Edward W. Glowienka
Leibniz's Metaphysics of Harmony
2016. 124 S., kt.
ISBN 978-3-515-11482-0

50. Nora Gädeke / Wenchao Li (Hg.)
Leibniz in Latenz
Überlieferungsbildung als Rezeption
(1716–1740)
2017. 262 S. mit 25 fbg. und 3 s/w-Abb., kt.
ISBN 978-3-515-11474-5

51. Thomas Leinkauf /
Stephan Meier-Oeser (Hg.)
Harmonie und Realität
Beiträge zur Philosophie des späten Leibniz
2017. 199 S., kt.
ISBN 978-3-515-11656-5

52. Wenchao Li (Hg.)
Leibniz and the European
Encounter with China
300 Years of "Discours sur la théologie
naturelle des Chinois"
2017. 295 S., kt.
ISBN 978-3-515-11733-3